A Ploughing People

A Ploughing People

The Farming Life Celebrated
Stories • Traditions • The Championships

Valerie Cox

HACHETTE
BOOKS
IRELAND

First published in Ireland in 2017 by HACHETTE BOOKS IRELAND

1

Cataloguing in Publication Data is available from the British Library.

ISBN 978 1 47365 945 2

Interior design and typeset by redrattledesign.com
Printed and bound in Germany by Mohn Media GmbH

Hachette Books Ireland policy is to use papers that are natural, renewable and recyclable
products and made from wood grown in sustainable forests. The logging and manufacturing
processes are expected to conform to the environmental regulations of the country of origin.

Hachette Books Ireland
8 Castlecourt Centre
Castleknock
Dublin 15
Ireland

A division of Hachette UK Ltd
Carmelite House, 50 Victoria Embankment, EC4Y 0DZ

www.hachettebooksireland.ie

Contents

To Brian

Introduction

'Ploughing is a religion, a way of life, an Irish mafia. It's not just digging up the earth, it's in your heart.' This is what I was told when I ventured among Ireland's ploughing families; from the mountains of Donegal to the plains of Kildare, these families place the National Ploughing Championships above all other arrangements, holidays, wedding dates – and possibly even funerals.

There is an intensity about them, a knuckle-grinding toughness, a love for their horses and their big shiny tractors, and a warmth that fires the embers of what it is to be Irish, to be part of rural Ireland. It's a world where there are always hot scones in the Aga, a chat with the postman, a smile for the stranger, the loan of a bale of hay or a rick of turf, a soft word for an elderly neighbour, a lift to mass or a clutch of eggs hatching away in the shed.

'Ploughing is a religion, a way of life, an Irish mafia. It's not just digging up the earth, it's in your heart.'

Despite the march of time, the vagaries of technology and the politics of the moment, for three days every September, the farmers of Ireland gather to compete against one another in the National Ploughing Championships. There are dynasties of ploughmen and women, people who learned the skills from their grandfathers, and their memories are sharp. They never forget the plot of scrub they were dealt at a match fifty years ago or the judge's decision that didn't go their way or the ploughman who managed to defer his turn until the frost had melted and the horses were no longer skidding about. They remember exactly where they'd lined up the three bottles of porter on the headland or spent half the night before making the sandwiches and packing the poitín into their wellington boots.

Today, the ploughing has expanded to the size of a small town – in 2016, there were 1,700 exhibitors and 300,000 visitors on the site at Tullamore, County Offaly. You could still buy a tractor, of course, but you could also join the priesthood, learn Irish, plant a forest, become a sheep farmer or join a political party. And speaking of politics, all politics is rural when it comes to the ploughing. You'll see TDs and senators posing on tractors, bonding with their rural constituents and standing ankle-deep in the mud in their shiny new wellies for the party photo for the local newspaper.

The tradition began in 1938, when President Éamon de Valera, a close friend of Championship founder J.J. Bergin, was photographed at the event. It wasn't quite the happy maidens dancing at the crossroads that the great man had envisaged, but it was close. The president had never actually ploughed himself, something he had mentioned in the Dáil during the debate on the Anglo-Irish Treaty in December 1921.

I lived in a labourer's cottage but the tenant in his way could be regarded as a small farmer. From my earliest days, I participated in every operation that takes place on a farm … I did not learn how to plough but, until I was sixteen years of age, there was no farm work from the spancelling of a goat and milking of a cow that I had not to deal with. I cleaned out the cow houses. I followed the tumbler rake. I took my place on the cart and filled the load of hay. I took milk to the creamery. I harnessed the donkey, the jennet and the horse.

For over eighty years, rural Ireland has come together each autumn at the National Ploughing Championships. It's where the people of the countryside meet to display their skills and meet their friends. Entire families train for these championships for the best part of a year.

This book celebrates the world of the plough, the ploughmen and women, the people who train the horses and shine the harness and make the tea, who spend their weekends at local competitions all over the country and for whom the real achievement is in keeping their ancient skill alive in their own communities. These are the people who till the soil in all weathers, the warm winds of summer and the biting hail and sleet of a winter morning.

In writing this book, I have travelled through towns and villages meeting some legendary ploughmen. I dropped in on a hooley in Dennis Kelleher's kitchen in Banteer, and I went to Abbeydorney in County Kerry, a village that has produced more champions than anywhere else, including the first Queen of the Plough, Anna Mai Donegan. There were no telephones in 1955 so she had to wait until she'd travelled back home from Athy in County Kildare to tell them she'd won!

Gerry King

I also met the sixteen-year-old boy from Louth who got a job as a bus conductor in London using his twenty-year-old brother's birth certificate. Today, seventy-four-year-old Gerry King, horse ploughman, has more All-Ireland horse ploughing titles than anyone else.

There's Joe Fahy from Galway, whose two greys, Paddy and Johnny, star in the Angelus on RTÉ. And Zwena McCullough, the first and only woman to plough against men in the Nationals, who was told by Thady Kelleher, 'You can't drink – and, if you can't drink, you can't plough!'

And, of course, there's the magic of J.J. Bergin, the man who invented the 'farmerette' and who ran the National Ploughing Association (NPA) for twenty-seven years until his death in 1958. His successor, the Kilkenny All-Ireland winning hurler

Sean O'Farrell, continued to grow the championships until his own passing in 1972, when the legend that is Anna May McHugh took the helm. She told Ryan Tubridy on *The Late Late Show* that at the first ploughing event she went to, 'I dressed myself up in high-heeled shoes and went out to a field but I came home minus one heel. I got sensible then!'

I finish my journey at a match in Roundwood, County Wicklow, where Kevin Doran taught me to plough my own furrow with my two new friends, Tom and Womble, his team of horses.

God speed the plough!

Anna May McHugh

1. The Beginnings

Ploughing is an ancient skill, it can be traced back to 4000 BC to the west of the Indus valley in what is now Pakistan. There is also evidence from 1200 BC from a painting depicting a yoke of horned cattle discovered in the tomb of Sennedjem in Ancient Egypt.

But we Irish can do better than that!

The late Sean McConnell, Agricultural Correspondent with *The Irish Times*, reminded readers that:

'Travel if you will to the Céide Fields in Mayo where 5,000 years ago, even before the Pyramids were built and the Roman Empire was established, Ireland's first farmers were tilling the soil.'

Ploughing goes back beyond written history in Ireland. Travel if you will to the Céide Fields in Mayo where 5,000 years ago, even before the Pyramids were built and the Roman Empire was established, Ireland's first farmers were tilling the soil. Professor Seamus Caulfield of UCD will show you a field close to his family home where there are tilling marks in the earth. These he believes may have been made by a spade but there is the possibility that they were made by a plough. He argues that a primitive plough may have been used because the fields, now covered by layers of bog, are all enclosed by stone walls to keep in farm animals which could have been used for ploughing.

And closer to our own times, competitive ploughing was a part of the life of workers in the large estates of the seventeenth century. At that time, wealthy landlords would pit their farm workers against rival estates, putting up generous prizes, in some cases as much as two months' wages or coveted tools like spades. There are records of a match held at Camolin Park in County Wexford on 20 October 1816. Five pounds was the prize money for a carpenter or ploughman who 'produced the best and cheapest plough made by himself and who contracted to supply the public with similar ploughs at the same price'.

However, it wasn't until 1931 that the first national ploughing match was held in Ireland.

> At that time a farmer would be rated by his own high standards. There was no machinery, everything was done by hand. So he had to have a good plough but he also had to have his hay ricks nicely shaped, like a Christmas cake, not scattered or anything wrong with them. He also had to have his farmyard tidy. But the main thing was he had to be a good ploughman and do his ploughing well.

The words there of journalist, author and public relations man Larry Sheedy, whose father was one of the founders of the National Ploughing Championship (NPC) and who first attended a ploughing match when he covered the 1954 World Championships in Killarney while reporting for the *Irish Farmers Journal*.

In 1931, Ireland was still suffering the impact of the War of Independence and an economic depression. The country's wheat average had declined to one of the lowest points since records were kept and the then Minister for Agriculture Paddy

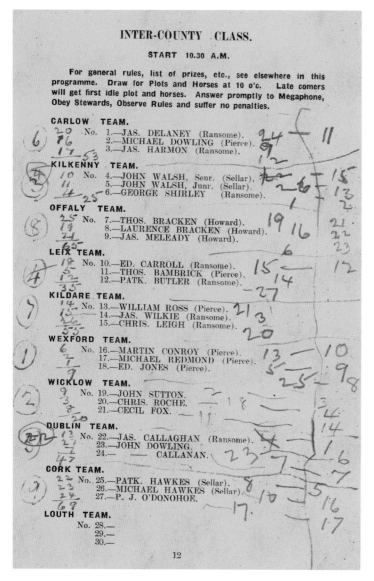

INTER-COUNTY CLASS.

START 10.30 A.M.

For general rules, list of prizes, etc., see elsewhere in this programme. Draw for Plots and Horses at 10 o'c. Late comers will get first idle plot and horses. Answer promptly to Megaphone, Obey Stewards, Observe Rules and suffer no penalties.

CARLOW TEAM.
No. 1.—JAS. DELANEY (Ransome).
2.—MICHAEL DOWLING (Pierce).
3.—JAS. HARMON (Ransome).

KILKENNY TEAM.
No. 4.—JOHN WALSH, Senr. (Sellar).
5.—JOHN WALSH, Junr. (Sellar).
6.—GEORGE SHIRLEY (Ransome).

OFFALY TEAM.
No. 7.—THOS. BRACKEN (Howard).
8.—LAURENCE BRACKEN (Howard).
9.—JAS. MELEADY (Howard).

LEIX TEAM.
No. 10.—ED. CARROLL (Ransome).
11.—THOS. BAMBRICK (Pierce).
12.—PATK. BUTLER (Ransome).

KILDARE TEAM.
No. 13.—WILLIAM ROSS (Pierce).
14.—JAS. WILKIE (Ransome).
15.—CHRIS. LEIGH (Ransome).

WEXFORD TEAM.
No. 16.—MARTIN CONROY (Pierce).
17.—MICHAEL REDMOND (Pierce).
18.—ED. JONES (Pierce).

WICKLOW TEAM.
No. 19.—JOHN SUTTON.
20.—CHRIS. ROCHE.
21.—CECIL FOX.

DUBLIN TEAM.
No. 22.—JAS. CALLAGHAN (Ransome).
23.—JOHN DOWLING.
24.— — CALLANAN.

CORK TEAM.
No. 25.—PATK. HAWKES (Sellar).
26.—MICHAEL HAWKES (Sellar).
27.—P. J. O'DONOHOE.

LOUTH TEAM.
No. 28.—
29.—
30.—

12

J.J. Bergin's notes from the first inter-county ploughing match, 1931

Hogan was urging farmers to increase production with 'one more sow, one more cow and one more acre under the plough'.

So in 1931, two farmers, J.J. Bergin from Athy in County Kildare and Denis Allen from Gorey in County Wexford, instigated the first inter-county ploughing match. Allen was already a member of Dáil Éireann, having been elected in the September 1927 general election for the Wexford constituency.

This whetted the appetite of tillage farmers around the country and, on 16 February 1931, men from nine counties competed at Coursetown in Athy. The organising committee cited the main objective as bringing 'the message of good ploughing to all parts of the country and to provide farmers with a pleasant, friendly and appropriate place to meet and do business'.

The cost of running the event was nine pounds, three shillings and five pence. There was a Perpetual Challenge Cup for the event, presented by Estate Management and Supply Association

Ltd in Dublin, and a team trophy, the David Frame Cup. Wexford took the team trophy and Edward Jones of County Wexford was the first individual winner.

But this was only the beginning; Bergin and Allen pooled their ideas and their energy to keep the project going and that was the birth of the National Ploughing Association. Bergin ran the NPA for the next twenty-seven years while Allen remained in politics until his death in 1961.

Allen's son, Lorcan, took his father's seat in the 1961 general election at just twenty-one years of age in a constituency reduced from five to four seats, and subsequently served as Minister of State on two occasions at the Department of Agriculture. He says his father could never have envisaged how the NPA has grown, but that he would have been pleased that his own county of Wexford is still very much to the fore in taking home the ploughing titles.

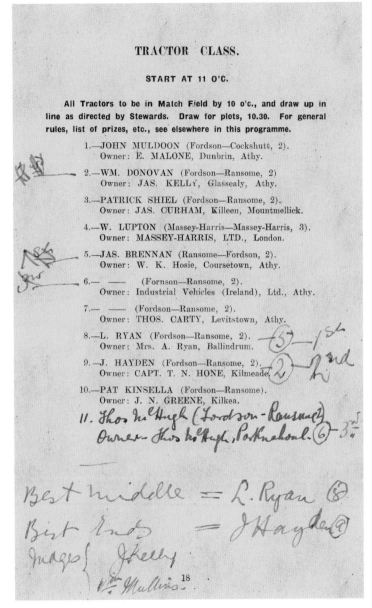

The competitors in the Tractor Class at the first National Ploughing Championship, 1931

Part of the attraction of that first competition may have been the natural competitiveness of the farmer. The poet Patrick Kavanagh wrote about the farmers of County Monaghan and the jealousy they felt for one another's ploughing prowess. According to Larry Sheedy,

> By all accounts, Kavanagh, himself a small farmer, wasn't reputed to be a great ploughman himself. The story goes that he used to get on his bicycle at the top of the road, get up a bit of speed, stand up on the pedals and he would be able to see over the hedges into the field to check his neighbours' ploughing. Kavanagh used to borrow a horse for his ploughing but he wasn't a patient man and he wasn't a tidy man and they were two facets of the good ploughman.

Sheedy gives the example of champion ploughman Martin Kehoe from County Wexford, as to what a ploughman should be. 'A huge strong man, a tug-of-war champion, a powerful man but he is powerful patient.'

'A huge strong man, a tug-of-war champion, a powerful man but he is powerful patient.'

And the late Sean McConnell of *The Irish Times* recalled his own early days when he wasn't allowed to plough close to the road where his handiwork could be seen by the neighbours. 'No one wanted crooked furrows which could be commented on by neighbours. It was nearly as bad as having an ugly wife or a wandering dog or no hay for the winter.'

There was a resilience to the ploughing which even the Second World War failed to deter. But the onset of war did slow down the progress of mechanisation. At an event in Ballinasloe in 1944, as many as ninety pairs of horses were borrowed in the area for the championships.

*J.J. Bergin and the Irish team enjoy a break
during the World Championships
in Canada, 1953*

But the demise of the horse was on the way and, in 1946, at the first post-war championships, the NPA introduced the first tractor competitions. Vigzol Lubricants put up the trophy which was won by John Halpin from County Wicklow.

J.J. Bergin died in March 1958 but not before he had left a wonderful legacy, including Ireland's involvement in the World Ploughing Organization of which he was vice-president. The first world contest was hosted by Canada in 1953 and the second came to Ireland, to Killarney in 1954 where ten countries competed.

Anna May McHugh

After Bergin's death, the NPA appointed the All-Ireland hurler Sean O'Farrell, a native of Kilkenny who had made his home in Wicklow, as managing director, a position he held until his own death in 1972. Larry Sheedy says he was an interesting contrast to J.J. Bergin – a quiet man, the opposite of volatile, known nationally as 'The greatest oil pourer on troubled boardroom waters of his period. Such a man, to remain so placid and in control, had to be made of stern, if concealed, iron.'

On the death of Sean O'Farrell, Anna May McHugh was appointed managing director of the NPA and she has remained at the helm ever since.

Sean O'Farrell

In the intervening years, she has attained the status of a legend, an icon so beloved of ploughmen that several ballads have been written in her honour.

Work on the next championships begins as darkness sets on the old one. In 2017, the Ploughing returned for a second year to Screggan, Tullamore in County Offaly. In the springtime I met up with P.J. Lynam, the NPA's National Chairman and he told me there were seventeen different farmers involved in providing the 700-acre site, the main one being Joe Grogan. The biggest development is the infrastructure, having enough water and electricity and toilets. 'The very same as providing a serviced building site for fifty houses,' he says.

Irish Examiner journalist Ray Ryan says the Ploughing is everything to rural Ireland.

It is one of those events that identifies with the people; for some reason people feel they have to go to the ploughing and it's not just people in rural Ireland nowadays, urban people come in their droves. And it always amazes me how people will head off into the countryside for a couple of days at the beginning of winter, cope with all the issues and come back again year after year after year.

'People meet each other, it's the end of the harvest and they might not meet again for twelve months.'

And Ray Ryan says his colleague Carrie Acheson of the *Examiner* always maintained it was the friendship that kept

people coming back: 'People meet each other, it's the end of the harvest and they might not meet again for twelve months. It's like a large family and there is never any trouble. There's still a kind of decency in rural Ireland.'

Despite recessions and technology, a wave of EU regulations, complicated grant systems and farmers' children gravitating towards a new lifestyle in urban Ireland, the heart of rural Ireland still beats as strong as ever. Farmers still love their land, their horses, their machinery and a world that still exists in the dawn rise on a hill, a brush with nature or a good price at the mart.

Farmers still love their land . . . and a world that still exists in the dawn rise on a hill, a brush with nature or a good price at the mart.

PLOUGHING CONTESTS AT ATHY,

On a Field at COURSETOWN, kindly given by W. K. Hosie, Esq.,

MONDAY, 16th FEBRUARY, 1931.

GENERAL PURPOSES COMMITTEE:

A. Reeves	Capt. Redmond	Jas. Ashmore
J. N. Greene	C. W. Taylor	Thos. Carty
E. F. Minch	J. Melrose	Hugh Kane
P. P. Doyle	W. K. Hosie	G. Mullins
J. Gracie	Jas. Kelly	Hugh Cogan
Capt. Hosie	R. Anderson	Capt. Webb
F. White	W. Duncan	E. J. Fagan
A. L. Spiers	P. Dooley, U.D.C.	J. C. Yates
J. Flynn	J. J. Keegan	G. W. Henderson
W. Cox	M. Malone	T. Ryan
John Owens	Ted Fennin	P. Kehoe
D. C. Greene	Jas. Duthie	J. J. Bergin

JUDGES:

For Tractor Work:—W. MULLINS, Goresbridge
JOHN KELLY, A.R.C.Sc.I., Carlow

For Horse Ploughs:—W. G. FOX, Slane, Co. Meath
RICHARD RINGWOOD, Shinrone, Offaly
W. McWILLIAM, Durrow Leix
E. BLACK, Glenealy, Wicklow
HENRY J. FOX, Kilternan, Dublin.

STEWARDS:

Competitors' Numbers and Plot Numbers :—Capt. Redmond.
Horses for Visitors :—J. C. Yates, Ted Fennin, C. W. Taylor.
Draws for Plots and Horses :—C. W. Taylor, A. Reeves, J. Crowley, A.R.C.Sc.I.

Tractors :—Capt. Hosie, E. J. Fagan, F. White.
Plot Surveyors :—J. Crowley, A.R.C.Sc.I., F. White.
Luncheons :—Jas. Duthie.
Starter :—C. W. Taylor.

Plot Stewards :—Nos. 1 to 4, T. Ryan; 5 to 8, E. F. Minch; 9 to 12, J. Gracie; 13 to 16, A. L. Spiers; 17 to 20, J. Melrose; 21 to 24, R. Anderson, Junr.; 25 to 28, W. Duncan; 29 to 32, J. Keegan; 33 to 36, Jas. Ashmore; 37 to 40, G. Mullins; 41 to 44, J. C. Yates.

Plot Superintendents :—A. Reeves, C. W. Taylor, Jas. Duthie, Capt. Hosie.
Press :—J. J. Bergin.

PUBLIC NOTICE :—The Committee will not be responsible for accidents to persons or property arising out of this undertaking, and all Competitors, Officials, and other persons attending, do so at their own risk and responsibility.

D. C. GREENE, Chairman of Committee.
JAS. DUTHIE, Treasurer.
J. J. BERGIN, Hon. Sec.

PROGRAMMES : : : : **3d. EACH.**

2. The Founding Fathers

J.J. Bergin

John James Bergin was the man who invented the farmerettes, who harvested corn on Christmas Eve and who hooked up a horse and plough and drove it through the town of Athy scattering a Cumann na nGaedheal meeting.

I'm visiting the Bergin home farm just outside Athy in County Kildare, which goes back generations and which is just a stone's throw from the original field where the first ploughing match took place in 1931.

'It's on the Stradbally Road,' says Mark. 'It's a town now, known as Coursetown, and it's still a farm but it's not tilled any more, a number of years ago the farm was leased to a dairy operation.'

I'm here to meet J.J.'s grandsons, brothers Mark and Andrew Bergin, who clearly remember their first visits to the Ploughing Championships as schoolboys. 'It was wet, it was miserable but it was a day out and we were obviously delighted we got a day off school!'

J.J. Bergin with President Sean T. O'Ceallaigh

But our talk turns back to their grandfather, a man who was a farmer, a politician of a kind and the assistant county engineer for Kildare. He was also on the Kildare Education Board and the Kildare Health Board, and was one of the guardians of the local hospital. He ran the local boxing club and was the leading light in the pipe and drum band; he was also a playwright and broadcaster as well as a 'terrible' poet. When he was young, he and his brother Andrew were very successful racing cyclists – indeed, the *Sydney Morning Herald* of 14 August 1905 noted Andrew's participation in the Pioneer Motor Cycle Club contest.

It is an amazing CV, but I'm interested in talking about the ploughing! Mark and Andrew have followed their grandfather and their father Ivan into the field of engineering and they have both ploughed, but not competitively, although Andrew had been out ploughing himself a few days before we met.

Mark and Andrew Bergin

'Of all of the agricultural operations, ploughing would have been the last thing we were let do, in the sense that it is precision cultivation, and you'd probably learn more about the settings of the plough,' says Mark.

But growing up, within the family, 'within the confines of conversation with our father, we would have been aware obviously of the history of the association and the nature of competitive ploughing and how it came about, the embryonic stages of it back in the twenties and thirties'.

J.J.'s grandsons are still in awe of the iconic spectacle that the ploughing has become. As Mark says:

> The Ploughing is fantastic, it's a celebration of agriculture, it's not just for the country person, it's all that's good about Irish agriculture. It embraces tradition, technology in every

'The Ploughing is fantastic, it's a celebration of agriculture, it's not just for the country person, it's all that's good about Irish agriculture.'

form, animal technology, computer technology, it reflects the advancing world that we live in, in an agricultural context, and it's a great social occasion as well! It's an incredible logistic feat to create a village or a town in a green-field site and hats off to the people who are involved. On a personal basis, we are descendants of where it originated, but you couldn't but be proud to be associated with it.

They are also appreciative of the work done by their grandfather's successors, Sean O'Farrell and Anna May McHugh. 'It's a huge credit to the people who have been involved in the past and who are running it now,' says Mark. 'Most of the work done is voluntary and to maintain that level of voluntary activity at a national level with very busy, very capable people is an extraordinary achievement.'

Their father Ivan was a meticulous man who kept very comprehensive family records of all the skeins in J.J. Bergin's life – photographs, letters, newspaper clippings – although some of their grandfather's memorabilia from the early days of the Ploughing was loaned about fifty years ago and never returned. The current generation has the advantage of having an historian in the family, their eldest brother, John. 'We're essentially going to make sure that the stuff is properly catalogued and archived. In time, there may be some local, historical agricultural thing.'

'He was a bit of a character, a man ahead of his time.'

We start discussing the fracas their grandfather created when, in 1955, he suggested running a Farmerette Class for women to compete for the title of Queen of the Plough – 'He was a bit of a character, a man ahead of his time.'

He was a Fianna Fáil supporter but, as Mark explains:

He would have been originally the Small Farmers' Party back in the twenties, which morphed into Fianna Fáil. His modus operandi with Fianna Fáil was that, essentially, the government in question was the beef barons in County Meath and the agriculture policy was very much geared towards them and towards the bigger players, whereas in this area, which was populated by large areas of small tillage farms because this is tillage country, there was very little security in terms of pricing, supply contracts and things like that. So going back to the early parts of the twentieth century, there would have been predecessors to the modern-day Irish Farmers' Association. I don't know if they ever really got traction on that.

'They didn't,' says Andrew, 'but it was all important. It was the first organisation of farmers – remember, farmers didn't even tend their own land. It was dodgy, people could be challenged on ownership of stuff. It was not too far off the Land League, and it was quite an uncertain time. To actually activate people and organise them was revolutionary, and it was frowned upon – farmers weren't supposed to do this because the merchants in trade, the millers, etc were in control. This was a powerful merchant class that you were standing up to and you were really getting above your station when you did that.'

Mark points out that J.J. did run in a general election in the twenties but Andrew says, 'It could have been as much a stunt as anything.'

'I suspect it was!' his brother agrees. 'I think if he really had ambitions to be a parliamentary politician, he would have been – he was that kind of man. He was his own man, a bit of a maverick character in some respect, hugely intelligent, quite highly educated for the time, he was an engineer as well as a farmer.'

J.J. Bergin had taken correspondence courses to become a civil engineer and his son, Ivan, became a mechanical engineer.

'It runs in the family, I suppose. By default, I ended up in that area too,' says Mark, 'but we're all of a similar disposition. I mean Andrew is farming but he's highly mechanised.'

The extraordinary thing is that while J.J. pursued all his interests, he was also Assistant County Engineer for County Kildare. I point out that there couldn't have been much free time.

Andrew agrees:

> Our father wouldn't have sat us on his knee and spoken nostalgically of his father, I sometimes wish he was still here to ask him. Somebody as prolific as J.J. Bergin, there can sometimes be a price to pay within a family, you know if you spend thirty-six hours a day working on all sorts of other things, and he farmed extensively as well. I remember my father telling me as a young guy, we were coming down past Rathcoole, coming out of Dublin, and it was starting to snow, and he told me that his father had rented land in Rathcoole for crops but it was some terrible year back in the early fifties when the harvest was delayed and delayed. They had a combine of some sort, which would have been remarkable at the time, and he remembered on Christmas Eve driving the combine home from Rathcoole. To farm that distance away today would be very challenging, to do it that length of time ago, with an old combine of some sort and to still be trying to cut corn at that time on Christmas Eve was an extraordinary undertaking. Eventually, they pulled the plug and brought the machine home.

From a farming perspective, I wondered if you could actually cut corn on Christmas Eve.

'It wouldn't be great,' says Andrew. 'You'd be doing it out of pure dogged determination really.'

So rather than a dreamer or a man of ideas, I asked Mark if his grandfather was a very 'hands-on' farmer.

'He was, although I suspect, given all of his other activities, that a lot of his "hands-on" would have been pointing at things that other people were to do!'

'It's a talent that we have inherited,' says Andrew, and Mark agrees.

'Our grandfather was also a fair man for the publicity,' Mark tells me, recounting a story from J.J.'s brush with politics that has slipped into the folklore of the area.

It goes back to the election in 1932 and there were two competing public meetings in the town, Cumann na nGaedheal were in the front main square and Fianna Fáil, the new party that was vying for power, was in the back square, and these meetings apparently were very animated affairs – it wasn't a good meeting if there hadn't been a good punch-up amongst supporters.

But, again, we were only ten years out from under the yoke of the British rule, the country was finding its feet politically, and a lot of what Andrew referred to as the 'merchant classes' versus the man in the street, the worker, the small farmer, and people got very passionate about this.

In any event, our grandfather hooked up a horse and plough and drove it into the town, and came into the square, down one side of the square where the Cumann na nGaedheal meeting was being held, drove it across in front of the stage, where the speakers were in full voice. The crowd had to part a little bit, like the Red Sea, and he created such a commotion! And then he went down to the back square where the opposing Fianna Fáil meeting was being held and half of the crowd apparently followed him, a bit like the Pied Piper. Our father told us this story, but he had a certain amount of poetic licence in his stories and we always thought this was stretching it – but I've spoken to two people who were at the meeting who remembered the incident. One of them was a publican, Frank O'Brien, he would have been in our father's class at school, he died this April aged ninety-five.

But J.J. was primarily an engineer and the discipline and thoroughness and attention to detail of engineering was transferred into his farming.

'Not that he was trying to run it like a factory,' says Mark, 'but this is tillage country and it's also highly mechanised country and a lot of the development and advances in agricultural technology would have had their origins within this region. There were new machines coming in, new farming technologies, new mechanical technologies.'

And in the first decade of the twentieth century, J.J. ran his own business manufacturing machinery parts and he had the Bergin plough catalogue parts and that was at the tender age of twenty-five.

In the 1950s, J.J. added broadcasting to his list of accomplishments with a weekly programme from Radio Éireann, as it was at the time. He wrote a play for the station but the brothers don't know if it was actually broadcast. 'It was something about a travelling salesman and it's hilarious reading it.'

'At that time, people wrote poems and ballads,' says Mark. 'Sure, that was the nature of the culture, bear in mind there was very little in the form of phone communication and not every house had a radio, so you're talking about a country that was coming into a very turbulent political time, with major social sea changes and, in the midst of all of this, agriculture was going from almost subsistence to commercial. It was well into the 1950s before rural electrification hit these places, it's not that long ago! The ballads and poems, they were his own PR machine, it was the Twitter of the day.'

'The ballads and poems, they were his own PR machine, it was the Twitter of the day.'

The Song of the Plough

Turn down the green, O: Man who ploughs;
Guide thou the plough with sharpened share
Turn up the brown to sapphire skies;
Mankind on thee for bread relies.
Bright shines the sun and God looks down
On man, on beast on hill and town,
Then sow the seed in mellowed earth,
To harrow's sway and wild birds' mirth.

The joyful hum of threshing time,
And later drone as mills make flour –
Mankind gets bread; but what man thinks
It was your sweat that forged the links?
But, sure the world must bend its will
In every age to ploughmen's skill;
Then, O: Hurrah, you men who toil,
You're masters of the sullen soil.

Turn up the brown, O: Men who ploughs!
The waken'd earth to warming sun,
And give all men their daily bread,
Your work is God's for He had said
He'll bless your work – your plough-team too –
Reward is sure for what you do.
Then, O: Hurrah, Sons of the soil,
God speed the plough, God bless the toil.

J.J. Bergin was also a bit of a poet and composed his own ballad
'The Song of the Plough' to honour ploughmen everywhere

Next, we delve into J.J.'s treasure trove, of which his grandsons are justifiably proud.

'That is a patent application dated 22 July 1904, for a turnip thinner – "for thinning or spacing Young Turnips, Mangolds, Carrots and the like" – here's the actual drawings.

'This is what we were telling you about, his own machinery production business. That was on Meeting Lane I think, where Matt McHugh's was.'

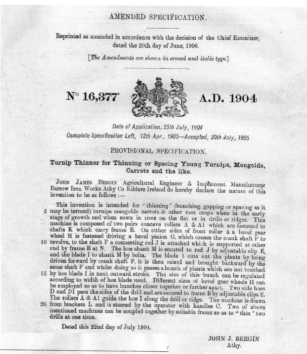

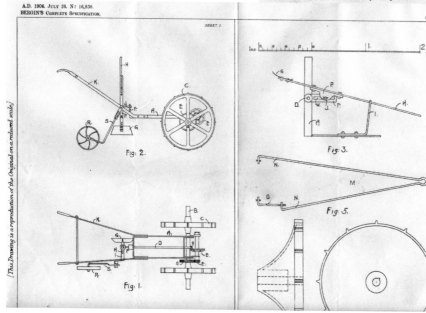

J.J. Bergin's patent for a turnip thinner, 1904

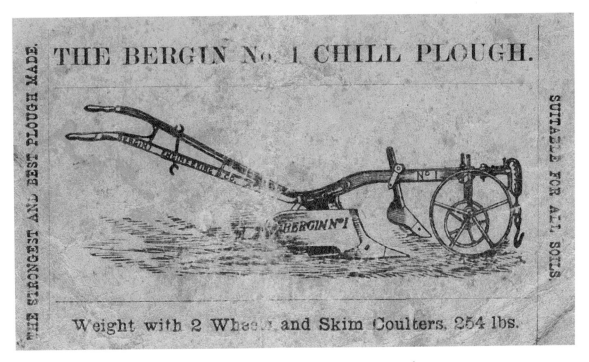

There's a catalogue for Bergin's Chill Plough Fittings offering The Bergin No. 1 Chill Plough 'suitable for all soils' and 'the strongest and best plough made'. And there are endorsements from farmers who had used his machinery.

J.J. Bergin's No. 1 Chill Plough

'I am well pleased with your No. 1 plough; it is very strong, and does good work in lea or stubble land. There is great wear in the fittings', wrote M Fennell from Kilerow, Athy, 3 February 1905.

And Jas. Byrne from Ballinteskin, Stradbally wrote on 17 March 1905 to say, 'I consider your No. 1 plough the best on the market; it did grand work in a very tough lea sod. I am sure you will have a lot of customers for ploughs from this district when they become known.'

On 14 September 1935, the *Irish Press* published a photograph of J.J. demonstrating a new harvesting machine which he had invented on his farm at Maybrook in Athy. The machine, it says, was known as a 'header and bagger', which cut and collected the grain.

But it wasn't all plain sailing for J.J. The *Kildare Observer* of 13 March 1915 reported that he had appeared at the Athy Petty Sessions to answer a charge that he had permitted an unlicensed driver under seventeen years of age to drive a car. He gave an undertaking that he would not allow it to happen again and had to pay the costs of 1 shilling and 1 pence.

There were always concerns among the farmers over seed supplies and there's a letter from The Maltings in Ballinacurra, County Cork, dated 26 February 1920, to say they regretted not being able to supply J.J. with three barrels of pedigree seed of any variety 'as they have all been engaged' and offering him an alternative strain. Andrew and Mark says this would have reflected the communications of the time. 'You almost had to beg suppliers and merchants to get what you wanted and if you were of a certain ilk or social class, you probably would get what you wanted and if you were a small croppy farmer, you were very much down the food chain.'

But in the thirties, the concerns of the tillage farmers – the ploughmen – over the price of grain boiled over when a deputation of them waited on Dr Ryan, Minister for Agriculture, at Government Buildings. The *Kildare Observer* of 23 September 1933 reported that Mr J.J. Bergin of Athy had told a press reporter that the deputation, of which he was a member, suggested that the minimum prices which would enable the farmers to carry on were: oats 14 shillings a barrel, barley 20 shillings and wheat 30 shillings. The pre-war price of wheat was 21 shillings while bread was sold at 5 pence per 4-pound loaf.

'Today, the price of wheat was much below 21 shillings, yet

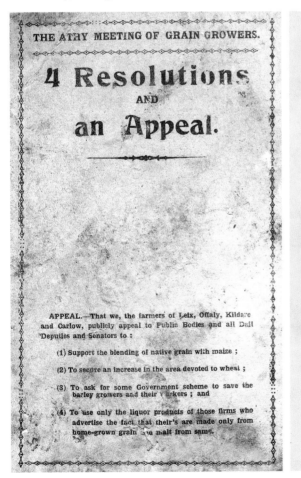
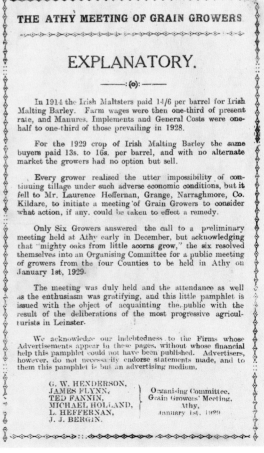

J.J. Bergin was a member of the organising committee of grain growers asking for an increase in the price being paid for wheat

the price of the 4-pound loaf was about 9 pence. The Minister,' said Mr Bergin, 'pointed out that increasing the price of wheat to 30 shillings would involve an additional penny on the 4-pound loaf. He agreed that the farmers' prices were now generally 10 per cent below pre-war prices, and while quite sympathetic to the point of view put forward by the deputation, said that the government's difficulties in this connection were many, owing to the various reactions which would arise from an increase in the price of primary products.'

It was with reluctance that I left the Bergin kitchen table and the vast encyclopaedia that was J.J. Bergin's contribution to the history of ploughing. He was not only a creative and inventive entrepreneur, but he had the ability to bring people along with him, to ensure that his dream of a National Ploughing Championships lived on long after his own demise.

Sean O'Farrell

On an early summer's day, I made my way to Enniskerry to meet Richard O'Farrell, nephew of Sean O'Farrell, the man who took over from J.J. Bergin as Managing Director of the NPA in 1958.

We're on Michael Keegan's farm because Richard wants to see the tractor and the plough that his uncle's old friend, Charlie Keegan, competed with to become world champion in 1964. Charlie's grandson, Michael, has painstakingly restored both and it's an exciting moment when we cross the farmyard into an insulated building and the covers come off the tractor! And there is the gleaming bright green D40 Deutz Kverneland alongside the shiny red plough that belonged to Ireland's first world champion ploughman.

When the founding Managing Director of the NPA, J.J. Bergin, died in March 1958, the NPA appointed Kilkenny All-Ireland hurling winner Sean O'Farrell to the position, a role he continued in until his own death in 1972.

Sean was born in 1909. 'In the same house and on the same farm that I was born in at Kilcurl, Knocktopher, in Kilkenny,' Richard tells me. 'My brother still lives there on the farm. Sean sent his only child, Sean Óg, to the same school with me as a boarder, the same school that he'd gone to for his Leaving Cert, St Kieran's College in Kilkenny. I knew Sean very well – I used to meet him every week for five years when he visited St Kieran's to see Sean Óg. It was there Sean fine-tuned his hurling skills, playing club GAA with Carrickshock because of its deep links with Kilcurl, which had been built up through cricket teams of the time and the Tithe War monster rally in Ballyhale with Daniel O'Connell. He went on to win an All-Ireland winner's medal with Kilkenny in 1933.'

However, Sean won his medal as a non-playing substitute, having injured his ankle while beating Galway in the All-Ireland semi-final.

'Sean was a strong and powerful figure and a great storyteller. We spoke privately for twelve hours straight just two weeks before he died suddenly when he told me his entire life story, including what he would do if he had life to live all over again.'

'Sean was a strong and powerful figure, and a great storyteller.'

Sean had two brothers, Richard, a Kilkenny farmer, and the Gaelic scholar Fr Pat O'Farrell. Richard remembers that: 'Uncle Fr Pat only spoke in Irish to our family as he saw the Gaelic language as the only key to our culture. He brought the O back into our family name in 1918, on the death of his father, whose headstone is in old pre-600 Gaelic script.'

He also had a sister, Mary Wallace, a well-known local historian who died in 1999 aged ninety-nine.

Sean emigrated to the UK before the Second World War but returned to Ireland and settled in County Wicklow where he married Lil Doyle, the owner of Lil Doyle's pub in Barndarrig. He lived there until his death in 1972 and was buried in the neighbouring cemetery where the oration at his funeral was given by Seán Ó Síocháin, the then-President of the GAA.

Always very involved in hurling in his adopted county, he was Chairman of the County Hurling Board and was very active in the campaign for the abolition of the Ban on GAA players playing non-Gaelic games, a rule that remained until 1971. He himself had played cricket for Surrey when he lived in England.

Sean is remembered as a very enthusiastic Director of the NPA. He brought in Esso as sponsors, followed by a host of major exhibitors. These were the main foundation stones of what was to follow in the NPA. During his tenure, in 1958, the NPA introduced a national bread-baking contest. Another achievement that pleased his fellow county men was bringing the All-Ireland to Kilkenny on three occasions, in 1959, 1964 and 1970.

Sean represented the NPA at the 8th World Championships in Rome in 1960, where he presented a block of Kilkenny marble as Ireland's contribution to the Cairn of Peace for the event. Then, in 1961, when the championships were held in Grignon near Paris, Sean met President Charles de Gaulle.

Richard tells the story of the photograph taken of the two men on that occasion. 'Sean was a big man, over 6 feet, but he was dwarfed by President de Gaulle who was 6 feet 5 inches tall. Nevertheless, he was very proud of that photo and displayed it on the wall of Lil Doyle's pub!'

1961 was also a very special year at home where the Ploughing Championships were being held in Killarney, County Kerry, and, much to the delight of the county, the Senior Horse Plough Class was won by a Kerryman, J.J. Egan, for the third time.

Telefís Éireann, the fledgling national television station, brought along an outside broadcasting unit. They filmed the event on 8 and 9 November for the first episode of *On the Land*. This was broadcast on New Year's Day 1962, the day after the station went on the air for the first time – quite a coup for the NPA. The programme

Sean O'Farrell shaking hands with Charles de Gaulle, 1961

had interviews with Sean and also with Mrs Grosvenor, the local event organiser.

The greatest success of Sean's tenure was in 1964 when the NPA sent two competitors to Fuchsenbigl, outside Vienna in Austria, where the late Charlie Keegan became the first Irishman to win a world title. And whatever about the buzz at international level, for Ireland it was an amazing achievement! Here was an Irish ploughman competing against the best in the world and taking home the title.

The following year, 1965, the NPA received another boost when Esso came on board as sponsor and introduced the

Esso Supreme Trophy, which is still presented to the Senior Conventional Champion, along with an NPA Trophy and a gold medal.

A year later, in County Wexford, a special competition was introduced for students from the agricultural colleges. Then, in 1969, there were more very popular innovations, as the Irish Countrywomen's Association gave cookery and craft demonstrations and the country markets were introduced.

Sean O'Farrell is remembered today as, arguably, the man who presided over the first golden age of the ploughing championships when Ireland's Charlie Keegan ploughed on the world stage and took home the first world title. This had an enormous impact on ploughing, the publicity and the celebrations encouraging other young people to come on board, a reaction that was to pay dividends on the home front even if the

Charlie Keegan arriving home with his trophy after being the first Irishman to win the World Ploughing Championships, 1964

international accolades paused for a while.

The NPA has been run since that first match in Kildare by three exceptionally strong leaders: J.J. Bergin, Sean O'Farrell and Anna May McHugh.

Writing this book, it is difficult to know where to place Anna May McHugh. As it happens, she is very much a part of the Founding Fathers (or

Anna May McHugh (left) and J.J. Bergin (right) at the Nationals, 1956

Mothers), having been recruited to the NPA in 1951 to work with J.J. Bergin. But she is, for most people, the figurehead of a modern NPA, an organisation she has brought from the realms of a relatively small ploughing match to the biggest agricultural show in Europe.

Having met this dynamic woman, having spoken to people all over the country who know her so well, I think her biggest attribute is her hands-on approach to her work. She has held the reins of the organisation since 1973 when she was appointed managing director but she has never allowed her status to interfere with her sense of community, of knowing her people and of keeping abreast of all things plough-related.

3. The Early Years

On 16 February 1931, ploughmen from nine counties came to Coursetown in Athy, County Kildare to pit their skills against one another in the first national ploughing contest. The Bergin family still have J.J's markings as well as the details of who had paid their 5 shilling entry fee.

At ten o'clock, the draw for plots and horses was held. Late arrivals had been warned in the competition programme that they would be given 'the first idle plot and horses', and all the

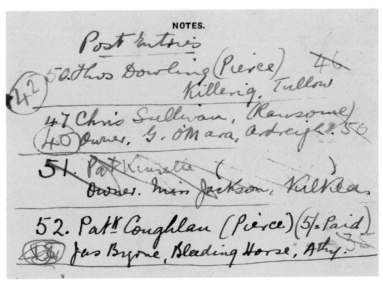

J.J. Bergin's notes showing late entries to the 1931 Championships and who had paid their entrance fee

competitors were warned to 'answer promptly to the Megaphone, to obey the stewards, to observe the rules and suffer no penalties'.

Forty-six horse ploughs were competing and as the men harnessed their horses and marked out their plots, they were probably unaware of events unfolding in the outside world – that one hundred people had drowned in the Pearl River south of Guangzhou when a steamer hit a rock and sank, that someone had slashed a Rembrandt painting in the Rijksmuseum or even that the original *Dracula* movie, starring Bela Lugosi as the titular vampire, was on its way into Irish cinemas, having been released two days earlier.

Denis Allen and J.J. Bergin were in charge, and Bergin was busy marking the score sheets for the various categories. There were classes for tractors as well as horses, and anyone who couldn't field a pair of horses could obtain them locally. Nine teams competed in the Inter-County Class – from Carlow, Kilkenny, Offaly, Laois, Kildare, Wexford, Wicklow, Dublin and Cork. County Louth had obviously planned to send a team as they are listed on the programme, but no names were posted on the day.

In the Championship of Ireland, there were seven entries, although four of them were late entries.

The top prize money was £12 in the Inter-County Contest and, in the Tractor Class, first prize was a barrel of oil valued at £7 for the owner, and £2 cash for the driver. Perhaps the strangest prize that first year was for 'the Married Competitor, with the greatest number in his family' – he won a 10-stone bag of flour!

On 17 February 1931, *The Irish Times* reported that:

> Wexford won the Championship of Ireland ... not only that,
> but one member of the Wexford team, Edward Jones, won the

Carter gold medal for the best work in the Inter-County Class, the ESMA Perpetual Cup, for the best ploughing in the field, a special prize for the best work done by an Irish plough, and a special for the best middle.

The paper had another headline: 'A Boy Winner' – in the Inter-County Class where 'a feature of special interest was the excellent work of James Ryan, a boy of about fourteen years of age who skilfully handled a Ransome plough belonging to Mr Joseph Fennelly, Athy, and won third prize'. The field itself was described as 'a strong loam, somewhat stiff and soapy'.

This may have been the first National Ploughing Championships, but there had been earlier, local contests that continued throughout the thirties.

PRIZE SCHEDULE

INTER-COUNTY CONTEST. For County Teams of Three Competitors. 1st Prize: The David Frame Perpetual Challenge Cup. Presented by D. Frame, Esq., Cooke Abbey, Bray; also, Cash Prize of £12. 2nd Prize, Cash Prize of £9. 3rd Prize, Cash Prize of £6. No third prize will be allowed unless five counties compete.

CHAMPIONSHIP OF IRELAND. The individual who, in the opinion of the Judges, does the best ploughing in the field. The ESMA Perpetual Challenge Cup, presented by the Estate Management and Supply Association, Ltd., Millicent, Sallins, Co. Kildare. Also, a Cash Prize of £5, and a Certificate.

LOCAL CLASS (Confined to Co. Kildare. For ploughmen who are not competing in the Inter-County Contest, and who have not won a first or second prize at any recognised ploughing match during the years 1928, 1929, or 1930.. First Prize, £3; Second, £2. Third, £1.

TRACTOR PLOUGH CLASS. For Tractor Ploughs, open to all. First Prize, Barrel Oil value £7 0s. 0d. to Owner, and £2 cash to Driver; Second Prize, 40 gallons Paraffin, value £2 5s., to Owner, £1 to Driver; Third Prize, Drum Oil, value £1 5s. to Owner, 10/- to Driver. Entry Fee, 5/- Above Prizes presented by Industral Vehicles (Ireland), Ltd., Athy. Special Prizes for Drivers: For best middle, 10/-; for best ends, 10/-

SPECIAL PRIZES.

For Competitor doing best work in Inter-County Class, a Gold Medal, value £5, presented by Messrs. James Carter and Co., Seed Growers, Rayne's Park, London.

For best work done by an Irish-made plough, Cash Prize £1, presented by Messrs. Cogan and Byrne, Auctioneers, Ballytore.

For best work done with a Howard Plough, Cash Prize, £1 1s. 0d., presented by Messrs. J. & F. Howard, Ltd.

For best middle in field, 10/-

For best Middle in Local Class, 10/-

For best Ends, confined to the Local Class, 10/-

For best turn-out of horses and harness, £1.

For the Married Competitor, with the greatest number in family, a 10st. Bag of Flour, presented by Mr. J. Gracie, Kilmeade, Athy.

The Cash Prizes, except where otherwise stated, have been presented by the Committee from voluntary subscriptions received, from Entry Fees, etc. The Department of Agriculture and the Kildare County Committee of Agriculture have contributed £5 from the Joint Fund as a subsidy to these competitions.

For best work done with a Ransome Plough Cash £1 presented by mess Ransomes Ipswich.

Prize schedule from the National Ploughing Championships 1931

Timoleague in County Cork, which celebrated its eightieth anniversary of the Ploughing in 2015, was one of the first to document a local match. Writing in his book, *80 Years of Timoleague Ploughing*, John Sexton says, 'It's a far cry from that January day in 1936, when the members of an agricultural class on the advice of their adviser, Jerry O'Donovan, decided on holding a ploughing match as a project.'

In her introduction to John Sexton's memoir, Anna May McHugh remembered some of the local men, including Larry Sexton, John's father.

> Larry was a man of great faith. I can remember whenever he was marking out the fields for the National event, he always stopped to say his Angelus and indeed an extra Hail Mary which was added to cool off some of his workmates.

And then there was Dan Connolly, best known for his sense of humour. 'Whenever there was any argument at meetings, Dan would always say, "Thank God to be here and well."'

But that first match in Timoleague was a day steeped in history, and for reasons other than the ploughing. Fr Dan Hourihane, a curate in Clogagh in 1936 and parish priest of Timoleague in 1970, spoke at a prize-giving following the 1970 ploughing match and recalled being invited to the first match in 1936 but, listening to the wireless during lunch-time, the news broke that King of England George V had died and, he says, 'the world stopped'.

His immediate thought was that the ploughing match would be called off, but, to fulfil his invitation, he decided to cycle west to Kilmalooda and, to his amazement, found the field full of people and ploughmen, working away 'as if the king had

Competitors in an early tractor ploughing match, Athy, 1929

never died'. Another parish priest, Fr Pat Hickey, later wrote, 'Those who told the poet Patrick Kavanagh that the plough was eternal were right. It is not disturbed by such little things as the coming and going of kings.'

(George V actually died close to midnight on 20 January but it may have taken a couple of days for the news to filter through. At any rate it didn't interfere with the ploughing.)

In his seventieth anniversary book, *70 Years A-Growing: A History of Timoleague Ploughing and Ancillary Events*, John Sexton wrote that he has a programme for a ploughing match held in Skibbereen in January 1896.

We do not know how many matches were held before this but what we do know is that the rules as laid out on the programme are almost identical to those adopted by the NPA thirty-five years later. Another report, which has come to our notice, is of a ploughing match held in Kilbrittain in the early 1920s on the O'Brien farm at Riversdale. Incidentally, when the Kilbrittain Ploughing Association was revived in 1983, its first fixture was held on the same farm.

It's important to remember too that ploughing in the early twentieth century was done by horses, mechanisation in the form of tractors didn't arrive until the 1940s. As equine specialist Wendy Conlon says, 'In the early 1900s, horses were status symbols for small farmers. The period up to the 1940s was the heyday of the working horse. There appears to have been something in the Irish consciousness that, at times, considered the social status conferred by horse ownership to be more important than good economic sense. Among the small farmers, owning a horse was regarded as desirable, even when their few acres did not justify such a possession in terms of commercial viability.'

Throughout the thirties and forties, even though horses were still used for ploughing – in 1951, there were 420,700 working horses in Ireland – the process was becoming more mechanised. In the ten years to 1948, the number of tractors used on farms increased by nearly 700 per cent to 9,211. And this changed the daily lives of farmers completely.

The National Ploughing Championships have evolved greatly since 1931, and have reflected the changes in rural life.

In the early years, the top title was the Horse Class, today that has become a much smaller class with the top classes in Senior

Conventional Ploughing and Senior Reversible Ploughing, as these are the classes where the winners go on to compete at the World Championships.

There are Under-28 and Under-21 Classes, Two and Three Furrow Classes and Reversible Classes. There is an Under-40 Horse Plough Class as well as the Senior Horse Class. There are also Vintage Single Furrow and Two Furrow Classes, as well as Novice Classes – the J.J. O'Dwyer Perpetual Challenge Cup goes to the youngest competitor in the contest.

Then, there's the Farmerette Class where ploughwomen compete to be Queen of the Plough. The winner receives the NPA Silver Crown as well as the J.J. Ruane Perpetual Challenge Cup.

Getting to a ploughing match

The logistics involved in getting the horses to the ploughing matches has always been complicated. John Sexton wrote about a conversation he had with Patrick O'Donoghue, a ploughman from Kilmeen, as they celebrated the golden jubilee of the NPA in 1981.

A few days before the All-Ireland, he took his plough in a horse and butt to Ballineen railway station en route by train to Athy. And the day before the event, he cycled to Bandon to link up with the Hawkes brothers. After the ploughing, Bergin entertained all the ploughmen at his home in Marybrook in Athy.

Another great champion, the late Thady Kelleher, told of the problems he had in even getting a team of horses to compete with at the World Ploughing Championships in Horncastle in England in 1984. Thady, his brother Denis and John Joe O'Connor crossed over from Dublin by ferry. They met the man in charge of the horses on Monday evening and arranged to get the horses on Tuesday. But this did not happen. On the Wednesday, Thady was given a pair of 'show horses' that had never been worked. Then, he got an offer of a pair of horses on the Thursday which gave him Friday to practise before competing on the Saturday. According to his late sister, Sheila Foley, 'This meant he had only one day to familiarise himself and build up a relationship and a binding trust with the horses.' But despite the daunting challenge, Thady took one of his international titles at Horncastle that year.

It's hard to understand now how difficult it was in the early years for our ploughmen to compete abroad and how expensive it was for them. The World Ploughing Championships came to Killarney in 1954, but it was 1960 before Andrew Cullen from Wexford competed at Tor Mancina in Italy and again in Austria in 1964 and Norway in 1965. Since then, the ploughmen and their families have blazed a trail across the world from Canada to New Zealand and from the US to Australia.

Tossing a hay bale, National Ploughing Championships, 1939

4. The Odyssey

There are ploughing hubs all around Ireland in the towns and villages that have taken more than their fair share of the accolades over the years. I set off on my own ploughing odyssey, which took me to farms and farmers in eleven counties and to Northern Ireland, where I met the support teams, the young, the retired and the champions – the people whose memories are part of the great ploughing story.

Carlow

My first stop was in Garryhill in County Carlow at the home of John and Lil Tracey. Lil is the matriarch of a family of achievers. Her husband John, her son Eamonn and her grandson Sean are all champion ploughmen and her father-in-law Michael also competed in the National Championships. I arrive to a great welcome, a sit-down-at-the-table welcome, as the table heaves with a roast and home-grown potatoes and a gorgeous apple tart.

But first I have to hear the story of how she met John more than fifty years ago.

When we were going to mass on a Sunday, you know the way everyone keeps to the one seat. Well, we went into our seat and John's family were on the opposite side down one. We were both eighteen and I had my eye on him, I suppose! We knew one another well, I met him at a dance one night and I was with another fellow and yer man was giving me a couple of dances. The next night, I told yer man I wasn't going to the dance. But I did and I met John. He didn't let me down. And that was it – we got married a year and a half later in 1967. I don't know where the years go, they fly!

But to go back even further ... John's father Michael was born in 1916 and had a small 40-acre mixed farm. He was

Sean, John and Eamonn Tracey – three generations of a ploughing family

*John Tracey showing his friends and neighbours the trophy
he received for his success at the Ploughing*

a ploughman as well and competed in the early NPA events, taking third place in a Junior Class in the 1930s.

John remembers, 'I used to follow him around when I was a young fella. I remember when I was thirteen years old we had to borrow a tractor for the Carlow Ploughing Championships and we both ploughed on the same day. But with the excitement that morning, I had a headache, and I was afraid to say anything in case I wasn't allowed go. Then in the middle of it, I got sick and had to come home and the younger brother, the twelve-year-old, finished the plot. You wouldn't be allowed to do that today. Between us, we won the novice category! Anna May came to the presentation in Mount Leinster Lodge that night.'

John's parents were thrilled with their sons' performance and it carried on from there.

This was 1955 and Carlow was very active in ploughing with three clubs in the county, which is down to just one now.

John continued to compete, winning his first national title in 1966 in the Under-28s, and then going on to notch up seven senior national titles, the most recent in 2008. He has represented Ireland nine times at the World Ploughing Championships and was runner-up six times.

His first international outing was in 1972 in Mankato in Minnesota in the USA. The following year, he was on home turf in Wellingtonbridge, County Wexford, where he took the silver, as he did in 1974 in Helsinki, Finland, and six years later in Christchurch, New Zealand.

John Tracey with his medals from the World Championships in Slovenia

John was seventh in the world in 1997 in Geelong, Australia, and in 2002 he took the gold runner-up medal at Bellechasse in Switzerland. In 2005, he took the silver again in Prague in the Czech Republic and then in 2009, his last international appearance, in Moravske Toplice in Slovenia, he was second overall.

'I never won it outright,' he says, 'but I have gold medals for the individual sections that I won.'

You'll see a very definite gap between John's winning runs, which came about because of a World Ploughing Organization rule that said when you had represented your country three times, you couldn't compete again. Then,

John Tracey arriving home from New Zealand, 1980

happily, Irishman Michael Connolly became a board member for the WPO and pressed hard for a change and, in 1979, the rule was extended to four years. John believes he may have secured a world title if he had been allowed to continue longer.

Ploughing means everything to John. It's still a big part of his life and he continues to compete at the Nationals. 'Sure it's like hurling or football or golf, you know, it's a hobby and an interest. You have no business being involved unless you have the interest.'

Of John's five children, his son Eamonn was the only one to become involved in ploughing. Eamonn, now forty-eight, started ploughing when he was thirteen. He won a novice match at county level then and his first appearance in an All-Ireland was in Urlingford in 1986. Today, he is one of the most respected ploughmen in the country with ten senior titles, an

47

intermediate title and an Under-28 title, as well as winning the World Championships twice.

The interesting thing is that father and son now compete against one another. 'When I went senior I was up against my father and I still am – every day I go out I'm ploughing against him! We're on a balance now. In the early days, he was winning but now it's swung around.'

And the family achieved a remarkable first in 2016 when father, son and grandson Sean all competed at the Nationals, three generations of one family – a record!

So back to Eamonn's World Championships. He first competed in 1998 in Altheim in Germany where he came eighth overall. The following year, he went to Pomacle in France and took gold in the lea, and was fourth overall as well as fourth in the stubble. He competed again in 2004 and 2006 and then in 2007, when he was third overall in Kaunas, Lithuania. There was 2011 in Lindevad, Sweden, and in 2012 in Biograd na Moru in Croatia where Eamonn secured second overall and a long list of individual awards.

In 2013, Alberta, Canada saw more awards and then, in 2014 in Bordeaux, France, he literally struck gold as the Overall Conventional World Champion. The *Carlow Nationalist* said on 19 September, 'The victory was all the sweeter for Eamonn and the Tracey family, given their phenomenal history in the competition, and the fact that a world title had eluded them thus far.'

Eamonn won again in 2015 in Vestbo in Denmark and then came second overall in 2016 at York, England, behind Scotland's Andrew Mitchell Junior. But it was a good year for the Irish, who have always punched above their weight in international ploughing. Wexford man John Whelan claimed a bronze, and a delighted Anna Marie McHugh, who is also General Secretary of the World Ploughing Organization, said that, since winning the Nationals last year to qualify to compete in the Worlds, 'The men have been working extremely hard preparing and now all their efforts have paid off. They have made their country extremely proud. Congratulations and a very well done to them both and all of their families.'

She went on to say that Ireland has always been highly respected in terms of producing ploughmen to compete on the world stage. 'However, Eamonn and John are held in particularly high esteem not only in Ireland but across the international ploughing community.'

John Whelan with the trophy he won for coming first in the European Ploughing Championships in the Czech Republic, 2009

The Irish team was coached by Declan Buttle from County Wexford and sponsored by Kverneland.

For his part, John Whelan has competed in the World Ploughing Contest on eight occasions between 2005 and 2016, his highest achievement being second overall in the Reversible Class in Denmark in 2015.

Eamonn Tracey has very good memories of his international successes.

> When I went to the world contests first, the pressure was on. Martin Kehoe had been there before me and people knew my father and his skills. The whole family came with me. It's exciting, you're representing your country.

Eamonn Tracey with his golden plough, 2014

There is no prize money involved at world level but the 'golden plough' is the coveted trophy for the overall winner. The original golden plough was actually made of gold but it became so valuable that the World Ploughing Board couldn't afford to keep it insured. It's in a museum in New Zealand now and they made a replica at the Sellafield plant in the UK in 1988. It's made from brass and you get to keep it for a year.

The actual ploughing at world level is physically gruelling 'because the plots are very large, 100 metres long by 20 metres wide,' Eamonn says. 'In Ireland, we'd never plough plots that big and you have less time, twenty minutes to do an opening split, which is two runs of the field, and after that you have two hours and forty minutes – a total of three hours.'

In the World Championships in York in 2016 there were twenty-seven entries in the Conventional Class and thirty in the Reversible Class.

There are separate competitions for Conventional and Reversible ploughing. The distinction can be hard to explain to the non-ploughing person, but the basic difference is the plough itself. 'In conventional ploughing, it's a one-way plough and all the furrows are being turned to the right-hand side. In reversible ploughing, it's a 'turnover' plough and there are angles, called 'butts'. When you get to the end of the furrow, you can actually turn the plough over and it will turn the same furrows back left-handed. You have left-hand boards and right-hand boards.'

But John says there are also big changes in the tractors they use. 'The modern tractor has gone bigger and higher, and you have to sit more forward. You can't really see what you're doing.'

Eamonn took over the farm from his father in 2003 and now farms 80 acres of tillage and beef. He uses a tractor every day, but says if he wasn't doing competition ploughing, he'd have bigger tractors.

While I was with the Traceys, I learned the secret of why shop-bought potatoes don't last any length at all. Of course they grow their own potatoes, great fluffy balls with flavour. 'They don't keep because they're washed before they're sold – the clay keeps them fresh. Unwashed potatoes last longer, anything starts to go off when it's washed because you're taking away the protection, damaging the skin.'

Going to the ploughing has always been a family affair. As Lil says:

I used to bring the kids to all the matches, it was very exciting. I remember getting all the kids ready for the matches, the babies in buggies, they loved it, watching their daddy ploughing, and you'd have a great time because farm families are all very friendly. You build your life around it. Every Saturday and Sunday, we'll be gone to the local matches from eight in the morning to eight at night. Eamon takes up the story . . .

'When we go to our local match, the families put up a tent in the field. They'd have tea and tarts and scones for when everyone comes. The wives of the committee members bring stuff and bring sliced pans and ham into the tent and make sandwiches. There's a lovely atmosphere. Then, that evening, up to ten years ago, the very same women would leave the field and have to go to a hall somewhere and start cooking dinner for about 150 people, usually the nearest hall to the field or a school. It was usually stew and ham and we gave the two choices of vegetables, eat it or don't!'

'You build your life around it. Every Saturday and Sunday, we'll be gone to the local matches from eight in the morning to eight at night.'

Things have changed a bit now, there's still the social end of ploughing, the camaraderie, but these days they go to the nearest pub or restaurant for the evening celebrations.

'Rural Ireland has changed so much,' says Eamonn, 'even farming has changed a lot over the past twenty years. But the ploughing match hasn't, it is still very similar to what it was sixty years ago. It's in our DNA. If I had a euro for every time someone asked me, "What does it take to win a world championship?" I'd be a wealthy man. I can't say what it is I have, I don't know what makes me turn a screw or shorten a link that no one else can see. A lot of fiddling around with the machine I suppose.'

The choice of equipment is paramount in giving yourself the edge to win. 'Our ploughs are Kverneland ploughs from Norway, the most popular match plough in the world,' explains Eamonn. 'Until the 1990s, there were a good few Pierce ploughs, a popular match plough up to then. We're better at grassland ploughing in this country but there is a world style ploughing which has taken over now – square-cut ploughing – whereas a Pierce plough is a kind of salmon-back ploughing. It looks sweeter and better, but the world style was leaving the furrows up in dead squares. Sure that happened you in Wexford,' he says to his dad. 'You probably had the best-looking job on the field, but then the world judges didn't go for that style.'

There are apparently fashions in ploughing and it's very complicated.

THE WISE FARMER USES A
Pierce Plough.

"THE PLOUGH THAT MAKES PLOUGHING EASIER."

PIERCE PLOUGHS ARE LIGHT IN DRAUGHT, SMOOTH AND STEADY IN WORK, ALWAYS KEEN AND LIVELY, AND MAKE PLOUGHING A PLEASURE :-: :-: FOR BOTH MAN AND HORSE :-: :-:

DIGGING AND RIDGING PATTERNS.
FOR ONE OR TWO HORSES.

SEE THAT YOUR NEW PLOUGH IS A
PIERCE,
BECAUSE PIERCE MEANS PERFECTION

CATALOGUES AND PRICES FROM—

PHILIP PIERCE & CO., Ltd., WEXFORD.

An advert for a Pierce Plough from the 1930s

'I don't know what makes me turn a screw or shorten a link that no one else can see.'

Some years ago, in Ireland, if you did world style ploughing, you wouldn't win anything because the judges here were on an Irish

Pierce style. You'd be ploughing here with a Pierce plough to go to the world and then, when you got there, you'd be ploughing with a Kverneland style plough.

In the 1990s, they started shipping out the competitors' equipment to the World Championships venue. For the first few years, they would send out the competitors' ploughs and they'd be given a tractor at the venue, but today the whole lot goes, 'and when you think about it, we're sending everything to the other side of the world'. But John recalls that, back in 1980, 'I went out to New Zealand with nothing only the suitcase.'

'I went out to New Zealand with nothing only the suitcase.'

Each ploughman has an individual style and they carry out a lot of modifications on their own plough to make it special. 'I have done so much work on mine,' says Eamonn. 'My mould boards, discs, everything on my plough is not the way it came from Norway. It's like a Formula One car. But you have to remember that the machinery may have changed, but the actual method of ploughing is still the very same even if the standard has improved immensely. When you're farming, you're fairly tied to time with what you can do but then my wife is always saying, if there was a ploughing match on I'd make time to go to it.'

'It's my hobby,' says John, 'and I look forward to the weekends and getting out there while the man above keeps giving me health and strength.'

And they have great hopes for Eamonn's son Sean, who is already making his dad and his grandad proud on the field. He has been competing in the Nationals since he was sixteen and already has four Under-21 titles to his name. And, although conscious that he is taking the family into the fourth generation, he says he enjoys it 'like any sport – we all pull together'.

Oak Park

The Oak Park Farm Manager, John Hogan, begins by stressing the importance of the top few inches of soil right across the planet, the soil that provides us with food. 'If that's not looked after, and minded, we'll all starve.'

Oak Park – the Carlow headquarters of Teagasc – has played a huge part in the history of the National Ploughing Championships, having hosted the event on six occasions (three of them consecutively) as well as the World Championships in 1996.

I'm here to meet the people from Oak Park and first I have to drive under the legendary Grand Arch. Designed by William Morrison, it was actually where the gatekeeper lived until 1970.

Oak Park House

competitors in the ploughing. Zwena McCullough, the Cork ploughwoman who came second in the world competing against men, remembers arriving at Oak Park in 1996:

The men would always talk about the entrance archway and all I wanted to do was to go through it! I had never in my life ever ploughed land like that, it was absolutely fabulous and I loved it!

'The men would always talk about the entrance archway and all I wanted to do was to go through it!'

The Painestown Estate dates back to the early 1600s and was owned by the Cooke family until 1785 when it was sold to the Bruens. Then in 1957 it was sold to an English farming syndicate and, two years later, it was taken over by the Land Commission An Foras Talúntais which had just come into existence, was given 344 hectares, including Oak Park House, to be used as a Tillage Crops Research Centre. An Foras Talúntais amalgamated with ACOT (the National Advisory Service) to form Teagasc, and in 2006 Teagasc leased 50 hectares of forestry parkland to the people of Carlow, which was developed into a recreational community and amenity facility.

So that's where I am – in one of the most wonderful working environments you could have. We are surrounded by fields, groves of ancient trees, wonderful old farm buildings and even a ruined church and graveyard, not to mention a partially built temple.

On a more commercial note, Oak Park is also the home of the Rooster potato and they're very proud of it. Gerry Boyle, Managing Director of Teagasc, tells me that Harry Kehoe, the man who produced the Rooster, Cara and many other varieties, is now retired but is still world famous. Eric Donald, Head of PR, tells me, 'He called it the Rooster because, apparently, when Harry was visiting a farm where they were trialling it, the farmer was asking him what it was going to be called. They've a very red skin colour and sure the day they were having the discussion there was a rooster there that colour ... so simple, so I think it was decided there and then.'

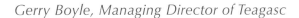

Gerry Boyle, Managing Director of Teagasc

'Oak Park has been chosen to host so many contests because of the space and geographic location,' says Gerry. 'It was probably one of the best sites for it.'

In 1990, there was considerable speculation that Oak Park might become the permanent home of the Ploughing Championships when, in a special supplement for the *Carlow Nationalist and Leinster Times*, it was reported that Gerry Baker, the then-Chairman of the NPA, said the association was interested in settling down to a permanent site, and that if a permanent site became a reality, Oak Park would be a forerunner. 'The reality is that our event is very large and it's more and more difficult to move into a green field every year,' he said. 'We have a chequebook that will not bounce and are prepared to buy if a suitable site comes on the market.'

But the NPC never bought a permanent site, and it continues to move around the country attracting huge numbers wherever it settles. Eric believes that one of the great strengths of the Ploughing is that moving it around gives them huge support from the communities that they move it to, though 'the traffic is nearly the most important consideration, you need the right soil, the right farm, but the traffic and the traffic management is critical'.

Oak Park, for its part, continues to carry out research into ploughing methods.

'The traffic is nearly the most important consideration, you need the right soil, the right farm, but the traffic and the traffic management is critical.'

We obviously plough here, but we do research on ploughing, on the best way to sow crops, because there are huge environmental concerns now. Compaction of soil is a big issue in relation to yields and management of water and climate change, there's a lot of research going on all over the world in relation to that. Ironically, there are environmental issues attached to ploughing now. Most of the research is now into less invasive methods of setting seeds, to minimise destruction of soil and compaction of soil. In many places, they don't plough at all now.

Teagasc demonstration on the compaction of soil

But on the question of whether ploughing could be on its way out, Gerry is adamant:

I don't think it will ever be on the way out, it's being managed more carefully, we're more conscious of the impact. A lot of the research is going into the tractor wheels and designing ones that minimise impact, putting tramlines in cereal crops and all this business. What's happened globally is interesting, the level of growth and productivity in cereal crops has flattened, and one of the reasons, people believe, is this compaction issue. We don't fully understand how the soil functions. We're looking now into fascinating research borrowed from humans – our guts are full of millions of microbes, and the same is true for the soil, so we're borrowing research from humans and applying it to soil. Some of our recent hires have been soil microbiologists, would you believe?

Eric explains that one of the things that farmers are really trying to do now is to manage the organic content in the soils. They're incorporating straws. 'Every couple of years you incorporate straw – for example wheat crops or oilseed or rapeseed – you chop it and you incorporate it back in, you also have the emergence of growing green cover crops, which are basically crops that are put in – rape, turnips and things like that – you put them in during September or August with a tractor. You sow them with a machine ... but what they do is they soak up the nitrogen that's available in the soil, it stays in the plant over the winter when the risk of it being lost out of the soil is the highest, and then you come along in spring and you either incorporate the green material back into the soil or else you graze it off with sheep or cattle and they do the recycling for you.'

Oak Park hosts many prestigious visitors and John tells the story of the time he escorted a group that included the Governor of the Central Bank around the estate. 'He was fascinated but there was one French guy and he wouldn't stand on the soil because he didn't want to dirty his shoes. The rest were laughing at him.'

There are so many interesting things to talk about including the plough and precision agriculture. 'We can measure very precisely the area of ground that we're covering, we can keep the tractor and trailer on very straight lines and that of course maximises the ploughing area and we can fit in all sorts of variables. They now have all sorts of instrumentation that they wouldn't have had such as GPS.'

'We basically went out with a tractor and cut out strips exactly every 10 or 12 metres instead of having to mark out with poles and everything.'

GPS works very well, so much so that on the most recent Open Day at Oak Park John says, 'We basically went out with a tractor and cut out strips exactly every 10 or 12 metres instead of having to mark out with poles and everything. We can go into a field, just like with the sat nav in your car, and pick up the feed and make lines at right angles and so you can say here's your field, we'll start sowing at this part. You can programme everything into the tractor to auto steer, depending on the width you want.'

But John makes it quite clear that such technical assistance is not allowed in competitive ploughing. 'Oh no! That's not allowed in competition!'

The championships held in Oak Park were so successful that it has been suggested they might return again. But Oak Park has become such a centre of excellence and experimentation that it would be a difficult proposition. 'If Anna May came here and said she was considering Oak Park,' says Gerry, 'the first person I'd go to would be John. John would tell me we'd have to shut down the trial programmes. John has to manage the plots, we do trialling all the time, experiments and controls. You'd have to shut down the operation, and it's not like an ordinary farm, you could have invested years in a research trial.'

I wondered if this meant Oak Park could never be used as a venue in the future, 'Never say never!' says Gerry. 'Anna May is a very persuasive woman ...'

Laois

Formerly Queen's County, the modern name of Laois is taken from the medieval kingdom of Loigis, and there is the sense of a kingdom about the place because Fallaghmore is the home and headquarters of the National Ploughing Association.

The successes of Laois ploughmen are scattered across many categories but everyone talks about Liam Rohan, who won the Senior Conventional Class in 1986 and who represented Ireland four times in the World Championships in 1982, 1985, 1987 and 1988, ploughing in Australia, Denmark, Austria and the USA. Together with Eileen Brennan and Tom Gowing, Liam was a founder member of the County Ploughing Association which celebrated its fiftieth anniversary in 2017. The three were all members of Macra na Tuaithe in 1967 when they decided to start the association.

In 2012, at the age of seventy-four, Liam Rohan died tragically in a farm accident and, two years later, his son and daughter-in-law, Brian and Norma, founded an organisation called Embrace to support the families of those who had died as a result of farm accidents. In 2015, there were eighteen fatalities on farms and in 2016 there were twenty-one.

Liam and his son John were working together on the farm when the accident happened. They were both knocked to the ground while they were dismantling a silage swarther and Liam was hit by a bolt on the side of his head. Liam was one of Ireland's best-known ploughmen but, as Brian says, 'To the nation, he was just another statistic in 2012 and he was going to be forgotten about, just like everyone else who had died in a farm accident.'

Initially, they didn't think Liam's injuries were too serious as he was still able to walk back to the house but he died in hospital three days later. The farming community rallied around and the Rohans had great support from their neighbours but, beyond that, there was a lack of official support for families bereaved by accidents.

Brian decided to hold a national remembrance service on the last Sunday in June to remember those who had died.

Embrace Farm ecumenical service at Abbeyleix, 25 June 2014

We encourage families to get in touch if they've lost somebody
or know someone who has died in a farm accident. Give us their
name and we'll call it out on the day and everyone is welcome
to come. Then the phone calls started to come in, tales of farm
families caught up in red tape after the death of a loved one,
bank accounts being frozen, mortgages on the farm, sometimes
no will left behind and the major problem of the cessation of
farm payments. We decided we had to do something to help
these people out so we got the wheels in motion and formed
Embrace in 2014. We have a Board of Directors in place now
and we're waiting to be registered as a charity. (Embrace can
be contacted at Annagrove House, Mountrath, County Laois
embrace.farm@gmail.com.)

One of Liam Rohan's friends was Michael McEvoy, who says Liam was the top ploughman in the county in his time. Michael is a bit cagey about his own age, which turns out to be a youthful seventy-one – 'But surely you're not going to write that down!' Michael farms in Coolderry, Ballacolla on a mixed farm. Growing up he was one of seven children and his father Patrick used a plough on the farm.

But it was a neighbour, Bill Meade, who inspired him to take up ploughing. When he was a small boy, Michael remembers being sent to bed on bright summer evenings and, instead of sleeping, he would look out his bedroom window and see his neighbour ploughing with a team of horses. 'I was always fascinated by how straight the ploughing was.' When he was fifteen, he started to plough with horses himself. 'But it was

'I was always fascinated by how straight the ploughing was.'

the sixties, a time when people were starting to get tractors and farming was starting to become mechanised, it was a transition period. Ploughing with horses was very slow. Next thing you'd know, a fella would come in the evening with a tractor and he'd do more in the evening than you'd have done all day with the horses.'

Some people kept a pair of horses for competitive ploughing, but they were being phased out in favour of the tractor. Even Bill Meade moved into tractors.

Bill and Liam bought two Ransome match ploughs in the early seventies and they started to encourage Michael to get involved in competition.

I'd be ploughing at home and Bill would say there's a ploughing match on in such and such a place and, like a fool, I went along! My first match was in Mountrath in 1968 and I didn't do well. I had no wheel on the plough, which shows how inexperienced we were. We lost at the match without a wheel – that was a bit of a disaster! I didn't have a match plough, it was the everyday plough from home, so I had a bit of a stop-start career in ploughing.

'I didn't have a match plough, it was the everyday plough from home, so I had a bit of a stop-start career in ploughing.'

But Michael went on to win the Three Furrow event in the Nationals in Waterford in 1983 and again in Ferns in Wexford in 1998. But, he says, he was 'probably a little bit unfortunate in another All Ireland' when he thought he should have won because although he says he had done the best job overall, he lost it on the skimming aspect. 'I was very disappointed, but what can you do but pack up your bags and qualify for the next one.'

The Nationals have been held in County Laois on six occasions, in 1943 it was in Lamberton Park in Portlaoise and Michael is delighted that it has been in Ballacolla 'right here beside me' in 1995, 2000 and 2002. In 2001, the Nationals were cancelled because of the risk of foot-and-mouth disease, instead a 'mini' match was held on David Lawlor's farm in Ballacolla to choose competitors to qualify for the World Championships. It was held in Ratheniska in 2013, 2014 and 2015.

Although Laois has held a series of very successful Nationals, Michael says the much-mooted idea of establishing a permanent site would not be a good one. 'People would get tired of the same venue. It would be a big attraction the first year, but then it would get tired as the event causes a fair amount of hassle for the landowners with car parks and so on.'

Michael and his wife Kathleen have five children but he says 'they have no interest at all in the ploughing – I'm the oddball in the family'.

I was given the same message by Pat Brandon. 'There is very little interest from the youth,' he tells me.

There's not enough being put into the youth to encourage them, we need to put a few pound back into the ploughing. It costs €5,000 for a plough now and a small tractor will cost you €15,000–€20,000. It's very hard for a young person to get set up and that's true for a number of other counties too, in Carlow for example there's very little youth only for the young Traceys. We need an incentive to get the young lads set up. I have consistently raised the issue. There was a plough for a prize for one or two years, but that fell by the wayside. Fellas like myself are retiring and leaving the tractor and plough to the young lads and that's how some of them get started.

Pat is the holder of twenty senior titles at county level in Laois and three titles at the Nationals in 1990, 1996 and 2009. He came third at the Five Nation event and represented Ireland at

Pat and Padraig Brandon

the Worlds in Denmark in 2001. He also coached the team that went to Limavady in Northern Ireland in 2004.

Pat farms 20 acres of tillage but in the real world 'I do a bit of everything' and that includes steelwork, making his own trailers, building his own house and roofing – indeed, our first interview took place while he was on a roof in blazing sunshine and I was on a mobile phone.

But Pat has another challenge these days, from his own son, Padraig, who has beaten his father a couple of times in the Senior Class at county level. 'I'm very proud of him,' says Pat. 'He won nearly all the youth matches and moved into the Under-21s.'

So it's time to meet Padraig, aged thirty-three, who tells me that 'ever since I was a young age all I wanted to do was drive tractors, anything to do with machinery'.

He first got really interested when he was eighteen and his father competed in the World Championships, but getting to matches around the country was a bit of a problem.

My father didn't have a big tractor, so I couldn't fit in to travel alongside him to get to the matches so I used to ring around and get a lift. I loved the old craic travelling in the tractor, we'd go from Ballylinan to Tipperary in four and a half hours, but we would pull in and let motorists by every now and again! I got great support from another ploughman, Fintan Keely. My father knew him so he gave me a hand yoking up a new plough. The following year, I said, 'I'm going to go by myself and whatever mistakes I make I'll be able to correct them.'

To date, Padraig holds six national titles from 2003, 2004, 2010, 2011, 2013 and 2015.

Padraig is a perfectionist and says, 'If I was ploughing the ground at home, I would hate anyone to drive by and I'm doing a terrible job.' He admits the enthusiasts are all ploughing 'snoopers, keeping an eye on one another's work and having the craic when it's below par'.

Eileen Brennan, co-founder of Laois Ploughing Association in 1967, three times a Queen of the Plough, Irish draught horse breeder and winner of the first brown-bread baking competition, is also a household name in County Laois.

Eileen competed in the Farmerette Class for seven years from 1959. It was her eldest brother J.J., who was also a competitor, who encouraged her to enter. '"Would you not try it someday?"' he said to me, and that was just it. In those times, if you were interested in something like that, the tractor people helped. Nuffield sent me down a tractor from Dublin because we dealt in Nuffield tractors then and Pierces of Wexford sent a plough up by train to Athy for me to practise until the event, and you brought it to the event. But of course it was good advertising for them!'

Eileen won the Queen of the Plough title on three occasions, in Killarney in 1961, at Athenry in County Galway in 1963, and at Kilkenny in 1964. Her reign was briefly interrupted in 1962 by Angela Galgey from Waterford, who took the title in Thurles. Eileen has a vivid memory of that first win. 'You were whipped away and had your hair done, then you had a fitting of the frock. Hillarys of Killarney, a huge drapery shop, suppled the gown and they had that in their window for the weeks before the ploughing match. It was a long pink gown.

'You were whipped away and had your hair done, then you had a fitting of the frock.'

Then you weren't to be seen again until you were to be crowned Queen of the Plough at midnight! It was very exciting.'

But the fun wasn't over. The Queen of the Plough led the St Patrick's Day Parade in Dublin on her tractor. 'We started off on St Stephen's Green and came down O'Connell Street, where you stopped for the salute, all the dignitaries were on the stand at the GPO. That was fantastic. Thousands of people lining the route, people brought their children to Dublin to see it.'

Eileen's daughter Elizabeth competed in the Farmerette Class on one occasion and her son Donal James was a competitor at national level in the Reversible Class. Eileen believes that if you become involved in something 'the touch never leaves you' so if somebody asks him, her son gives a hand to other ploughmen.

Eileen says she was always one of those people who took part in things but she sets limits. 'I would say, and I still aim at, five years. That should be enough for somebody to get to where they want to get; if not, it's time to bail out and go!' Eileen was one of three girls and five boys – and one of those sisters is Anna May McHugh.

Eileen went on to become interested in and to breed and judge Irish draught horses. She brings groups of people to 'GB' as she calls it, to the draught horse shows there. 'When you see a mare and you see her foal, it's good to leave a gap and come back again and see them again in a couple of years.'

She has three horses herself, one is being ridden, one is in foal and one is retired. Eileen explains that the Irish draught horse is the foundation stock of the Irish sports horse. 'It means that you breed an Irish draught filly foal for example, you bring it up and if you cross that with a thoroughbred horse, you have the

'Then you weren't to be seen again until you were to be crowned Queen of the Plough at midnight! It was very exciting.'

Eileen Brennan (right) with her granddaughters

half-bred, and that's the show jumper of today. Cross it again to the three-quarter and you have the event horse or the high-class show jumper. It's the foundation stock of all horses in Ireland.'

Eileen has also been involved in a local pony club for more years than she wants to admit to!

Then, there's the brown bread. At the World Ploughing Championships in Wellingtonbridge in Wexford in 1973, the competition was introduced and Eileen won. 'Last year Marty Morrissey interviewed me for RTÉ about the bread-baking and it was lovely, it was grand.' Today, she judges the competition and she also judges the best pair of Irish draught-type horses ploughing at the competition. 'I'm quite busy and I always say I'm on the phone if somebody wants me.' Eileen is a very positive woman whose parting words were, 'I am where there is laughter and joyfulness and happiness. I am there.'

Galway

'I met a man from Aughrim at a sale in Ballinasloe – Keith Martin is his name – and he had two beautiful greys. My wife Julia fell in love with them. We had a great set of old tackle and there's not many like it, bridles, collars, they were beautiful on the two greys. But the horses had never ploughed before. We went up to look at them and didn't it come snowing. But we tackled them up anyway, and we went driving around in the snow pulling a dray! They were walking too fast, but I felt they had great potential and I took them home. They settled in so well you'd swear they'd been ploughing every day.'

Julia and Joe Fahy, with their horses, Paddy and Johnny

I'm talking to Galway's champion horse ploughman Joe Fahy, and he tells me the two greys in question, Paddy and Johnny, are the team that appeared on RTÉ's Angelus for two years.

Joe and his brothers Michael and Padraic decided to introduce the new horses to the plough so they took them to the farm of the Franciscan monks at Currandulla, just across the fields from where Joe lives. 'They worked away and two days after that we took them to a ploughing match.'

Joe's father Pat was also a ploughman but 'he was always very unlucky, often beaten by just one point, although he did win a County Junior final in 1936. My father started with horses on the family farm. He came from Lisheadford and our

Joe and his niece Caroline competing in the Under-40 Horse Class, 2013

two next-door neighbours were within 20 yards of where we were brought up, we were all third cousins.'

Joe is also very proud of his sister Philomena, who came second one year in the Queen of the Plough, and of his niece Caroline Lardner, who competed in the Under-40 against the men at national level on a number of occasions.

Joe himself competed in his first ploughing match in Claregalway when he was sixteen. 'I was hardly strong enough for the job that time, but the next year I competed in the county Under-21 competition and didn't I win it. And I won the following year as well, and I was beaten the third year! Then I came back and won it two years in a row. In 1964, I entered the Nationals and it took me up to 1972 to win the senior All-Ireland. I can be awful thankful.'

'I was hardly strong enough for the job that time, but the next year I competed in the county Under-21 competition and didn't I win it.'

Joe also took the senior title in 1994 and 2000.

His horses then were Molly and Dolly, two black horses that he bred himself. But as the two mares were getting very old, Joe thought he might never have as good again. His winning streak with them started in 1990 and he is one of just two horsemen (the other is Gerry King) to have won All-Irelands in two different centuries. Molly and Dolly passed on aged twenty-six and twenty-seven. 'I had it packed in then, I decided I'm finished now, I'm enjoying myself and then, six years ago, didn't we meet up with this other fella with the two greys!'

Julia adds, 'Joe got out of the ploughing and this man came along and said, "Well, you're not, I have a team of horses and you're very welcome to them and I'll bring them back up from Cork", and that was six years ago.'

These horses were fairly old as well. 'They were making out that one of them was shoving up to forty and I was afraid of my life he'd die on me while I was looking after him! Julia always said if anything happened to these horses, what would we do? Then a neighbour said, "Haven't ye a fine field up there to dig a big hole for them?" The older horse did die, but only after he had gone home to his owner.'

The interesting thing was that the horses had belonged to Joe's old friend, the great Thady Kelleher.

'These horses were trained, they'd nearly talk to you.'

I knew him from the ploughing. These horses were trained, they'd nearly talk to you, and one is still ploughing in Galway. Often we were in the sheds at the ploughing, getting them ready for the match and my brother Paul would be going on with Thady saying, 'We'll beat you today!' Thady took such a liking to us!

These plough horses are very special, I've a great love for them, I was brought up with them! We ploughed all the way from Cork up to the North of Ireland, we know people and people know us just because of the ploughing.

However, it's Julia who remembers the military operation involved in getting the whole family off to the ploughing. 'The neighbours would bring the lorry and the kids – Julie, Patrick and Denise – they all loved it. I'm convinced the horses enjoyed it too! Joe's brother Padraic competed as well and he used to come and help Joe with the horses.'

But this is where the Irish mammy really comes into her own. Apart from getting the whole family ready, there's a full day of polishing the tackle with Brasso, washing and grooming the horses, plaiting their manes, painting the plough and making the

Joe, Michael and Padraic Fahy

sandwiches. It was always a family affair in the Fahy household and the neighbours got involved too.

'In the local competitions, it was tea and sandwiches from the back of a trailer, but the Nationals … that was a holiday! Bed and breakfast for a week not to mention the devilment that happened! Whole families rented houses and played tricks on one another, you had to make sure you kept your door locked!'

Joe is a bit wary around journalists ever since he met an unnamed woman reporter at the Ploughing. 'I gave her a hold of the horses and do you know what she wanted to do? This beats Banagher altogether! She wanted to get our team and another team and to put them chasing down the field with the ploughs and all after them! Crazy idea!'

Joe's horses had been on television before, but what we have to understand for this story is that the soil on the Fahy farm is very stony. 'It's alive with stones. We have awful bad land and, afterwards, a neighbour was down in the south of the country and he met someone who had seen the programme and he said, "Jaysus, haven't the Fahys awful bad land! How could they make a living in that sort of land!" And my neighbour said, "You didn't see the worst of it at all. Sure you don't think they were going to put the worst of it on television!"'

Joe speaks very affectionately of his mother, also called Julia, who, he says, was crippled with arthritis from the neck down for twenty-eight years. Despite being wheelchair-bound for thirteen years, she insisted on going to watch Joe compete in the ploughing. 'Just imagine!' says Julia. 'And she'd get so afraid that Joe wouldn't come first! She was sitting in the front of the car and she nearly came out of it when he won!'

Then there was Joe's grandmother, another Julia. 'She never thought she would live to see me getting an actual trophy,' he says, 'because it was just money prizes back then, no trophies. So didn't I buy a little cupeen in Woolworths of Galway for half a crown and I gave it to her when I won in the Under-21s. She thought she had the Sweep won!'

Joe stresses that his life in ploughing has been a family affair. 'There's no use saying anything but the truth, it was the family, brother, sisters, wives, every one of them had a hand in it and the likes of the neighbours who were involved. We could go away for a week to the Ploughing and tell the neighbours and they'd look after everything till we came back, great neighbours.'

Joe came to the scene after the legendary Thomas Reilly who won a junior title in the Horse Class at the Nationals in 1963 and went on to win a total of ten senior titles. Tom represented Ireland at the World Championships in Lincolnshire in England and was runner-up. They were good friends, living just five miles from one another and they often spent long winter evenings together plotting their ploughing adventures. 'If we won, we won! We didn't take it too seriously!' says Joe. And he says the greatest honour he had was when he walked up to the altar at his friend's funeral mass in 2002 carrying the sock of Tom's plough as the gifts were brought up.

'There's no use saying anything but the truth, it was the family, brother, sisters, wives, every one of them had a hand in it and the likes of the neighbours who were involved.'

Joe remembers how Tom used to plough together with his brother Johnny. 'They were partners in the ploughing and, when Johnny died, everyone thought that'd be the end of Tom and the ploughing. But instead, he surprised everyone and went on to win ten senior titles with the horses!'

But, according to Tom's son Gerard, his Uncle Johnny was 'the brains of the outfit'. Tom grew up on a small farm in Claregalway but when he was only twelve years old, his father died and he had to leave school to help his mother on the farm. There were three boys, Tom's twin and Johnny, his older brother. Tom got involved in ploughing through their local agricultural advisor in the forties and won the first of his twenty-eight county titles in 1945. This was the heyday of the horses with up to fifty teams competing regularly in the Galway area. But even though Tom qualified to take part in the National Championships on many occasions, he missed a lot of them because the event coincided with the beet harvest.

Tom and Johnny went to matches together and Tom was delighted that his brother got to see him win his first national title in 1978 before his death in 1980, still in his early sixties. But then Tom's career took off, winning another eight titles between 1981 and 1993. 'He was in his sixties then,' says Gerard. 'It was an amazing time to have so many wins. He used to say, "The older I get, the straighter I plough."'

Gerard Reilly, James Kilgarriff and Pat Quinn,
working the horse and plough

Tom's last success was in 1993 at a local match and he was still ploughing into the late nineties at competition level.

But Gerard has taken over where his father left off, winning three consecutive Under-40 titles in 1990, 1991 and 1992 and following that by winning the Senior Horse Class in 2014 and 2016.

Even though Tom died in 2002, he was around to see Gerard's early victories. 'I think he was more delighted with that than the wins he had himself. I think he was afraid that the skill would die with him. I had no choice but to get involved in ploughing and I was involved all my life – since I could walk I used to be with him.'

'I think he was afraid that the skill would die with him.'

He has his own team of horses, Duke and Blackie, which he bought as foals and broke and trained them himself. Tom's son-in-law James Kilgarriff is also a successful ploughman, having won the Special Horse Class in the Nationals in 2010 and 2012 and the Under-40 in 1999.

Wicklow

I'm in Enniskerry and there's a farmer offering to bring me up to a rock to see the marks left from a bonfire that burned fifty-three years ago to celebrate the achievement of Charlie Keegan. Robert Roe is the NPA Director for Wicklow and he still remembers the excitement of Charlie's return home to Enniskerry.

Two lorry loads of tyres were brought up by a local man, Dan Nolan, and I remember I wasn't able to roll the tyres but I was able to watch them being rolled up on the top of the rock, and they lit a bonfire and it burned for two days. And after all those years, the hairline cracks and red on the rock are still there to be seen.

Then, they had a marquee down in Charleville House. It turned into a fantastic year, every weekend there was a dinner dance because of Charlie, everyone wanted to honour him. I remember being at some of the dinner dances, I was only a kid, but I still remember. I was actually at the airport when he arrived back from Finland, and there were a lot of drinks on the tables but I wasn't allowed near them. I can still picture all the people waiting for him and the press were there in a big way. Charlie Keegan's name was worldwide. Robert's father was a close friend of Charlie's.

Going back to when my mother first got married to my father, they lived on an outside farm which adjoins this farm here, and she remembered Charlie Keegan coming

Charlie Keegan and the golden plough at one of the many gatherings to celebrate him becoming world champion, 1964

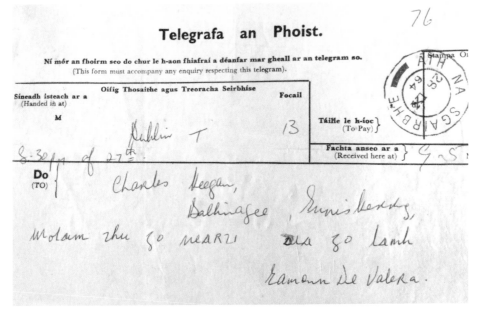

The telegram Eamon de Valera sent Charlie Keegan to congratulate him on his win

knocking on the door, looking for my father, because he wanted to do some practising for the ploughing match. And at that time, he would have had horses, and he would go down in the evening and put up a set, or a middle, to start off the ploughing.

Charlie was only twelve or thirteen when he started ploughing. His own father wouldn't have been a great ploughman, but he liked to see the ploughing. My father had a lot of experience, and, after that, my father used to judge in all the local matches. He did compete and he would have won at local level. But he never ploughed at national level.

Charlie Keegan went on to plough and to win at the nationals and then, in the 1950s, he moved from horses to tractors and

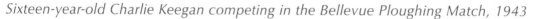

Sixteen-year-old Charlie Keegan competing in the Bellevue Ploughing Match, 1943

Robert's father 'sort of lost a bit of interest then because he wasn't that interested in the tractors'. Another local man, Jim Jones, he took an interest in Charlie then, 'and encouraged him' until he won the World Ploughing Championships in 1964.

I have my own little memory of Charlie here because I interviewed him surrounded by his trophies when I worked as a reporter with the *Wicklow People* back in the eighties.

Charlie had two sons, Stewart and Alan, both of whom predeceased him. Alan had also ploughed and came in fourth place in an All-Ireland Under-21. Alan's son Michael, who's now in this thirties, farms the family farm and has great memories of his grandfather. But he didn't realise what a celebrity he was until much later. 'My grandfather just happened to be a ploughman,' he says.

The Keegan family

Charlie Keegan controlling the plough

When I spoke to Michael, his wife Hannah had given birth to their second daughter two weeks earlier, but he couldn't remember the date. 'The lambing was coming to an end at the time and she was the last lamb!' he says diplomatically.

Michael's father had been ill from an early age and in fact died when he was only eighteen, so his grandfather became his mentor on the farm, teaching him how to plough as well as all the basics. Michael says his grandfather was an excellent farmer with a great variety of enterprises. Over the years, they had dairying, pigs, sheep, cattle, tillage, horticultural crops and even, in later years, a farmers' market. 'I worked quite a bit on the farm and I loved it. I never wanted to be anything else. It's a great life. There are days when it can be hard. Like any job, it has its ups and downs, but I love the life I have.'

Michael remembers Charlie as 'a very honest and hard-working man'. And hopes the same values have rubbed off on him. 'It was a very hard life for my grandfather's generation, everything was done by hand. There were no tractors in the early years, it was all horse ploughing and an acre a day would be a long slog, you might even have to change horses during the day! It was always wet and cold and even when the tractors came in there were no cabs on them. He had a hard life, whereas I can now plough ten to fifteen acres in a day without breaking a sweat!'

Although Michael wasn't even born when Charlie won his world title in 1964, his achievements live on in local memory. '"You're Charlie Keegan's grandson!" – I get that all the time. It was massive, you've no idea, it was like Ireland winning the World Cup, I think the only title Ireland had then was Ronnie Delaney in the Olympics.' Michael also points out that Ireland was an agricultural nation at the time so the win meant everything to the farming community.

The Restoration

Ahead of the fiftieth anniversary of his grandfather's title, Michael restored his tractor, the D40 Deutz Kverneland, the one he had ploughed with in Vienna, and it was displayed at the ploughing championships in Ratheniska, County Laois in 2014. It took him nine months to bring it back to pristine condition, but he had a lot of help collecting parts from scrapyards, online, from neighbours.

In the 1980s, Charlie had sold his tractor to a man called Derek Stringer. Derek then very kindly agreed to return it to

Michael so he could restore it. They came to an agreement whereby Michael found another vintage tractor to give to Derek in exchange for his grandfather's tractor.

It had been lying in the back of a shed. 'It was fairly rough, but it had been quite an expensive tractor for its time. It had become extremely popular after my grandfather's win and so people bought them. A neighbour of mine, Mick Hogan, had one and gave me a lot of parts for it.'

Then along came a businessman with an interest in vintage machinery, Tony Killarney, and he offered sponsorship, and Michael's wife Hannah began to publicise the project. Then the offers of help came in and a number of people came on board to get the restoration underway. Peter Jones, a machinery expert, and his father did the bulk of the restoration work on the tractor itself. Tom Murphy rebuilt the engine and Jim Larkin imported some of the bits they needed for the plough. Then Harry and John Williams got involved with their knowledge and contacts and John returned all the extra parts for the plough, mould boards, etc., which had actually been given to him by Charlie years earlier. Ecoblast in Kilcoole sandblasted the tractor free of charge and Kverneland Ireland came on board. And locals Robert Roe and Benny Doran gave their support so very much a successful community effort!

'He would have ploughed in the World Championships and, when all the hullabaloo was over, he came back here and did his farming just the same as if nothing had ever happened.'

The actual plough was a bigger problem because Charlie had cut it up for parts to repair other ploughs! 'It was just a plough to him, that's all it was. He would have ploughed in the World Championships and, when all the hullabaloo was over, he came back

here and did his farming just the same as if nothing had ever happened. He was very proud of it, but would never have been the kind of person who would have bragged about it. But the frame he dismantled was as scarce as hens' teeth, only about fifty were ever imported into Ireland. Eventually I found parts in the UK but I had to remake some of them.'

The Enniskerry Ploughing Association

The Enniskerry Ploughing Association is, according to Robert Roe, the oldest committee in Wicklow. 'Rathdown say they're the oldest but Enniskerry has been going for well in excess of 120 years.

> It all started with Lord Powerscourt. All the gentry had their own workmen and when they'd be socialising together, they'd all claim to have the best ploughman! So they got together one night and decided to settle it once and for all by having a competition among their workmen. So it came from the very top down and they put up fantastic prizes – £1 10 shillings, which was an awful lot of money – wages were only a shilling a day – so it was a couple of months' wages in one day. Men used to go mad to get out to try and get that prize. The competitions used to be held in the Powerscourt estate in the early days, and then it spread out to the local farms. That's why it's called the Powerscourt Ploughing Society.

We chat about the past champions, especially the Dowse brothers, Edward, Joe and the late Henry. Edward, who is in his eighties, still judges at the Roundwood show and is Chairman of the County Ploughing Society. He won the Under-21 Class at national level and he also won in a Tractor Class at Ardfert, County Kerry. 'I remember it well,' says Robert. 'Not only the same year – 1984 – *but the same day* that all the arms were found on the fishing boat, the *Marita Ann*!'

Seven tons of military equipment were confiscated on that occasion and the crew were arrested off the Skelligs and the fishing vessel was escorted to shore by two navy vessels.

'Muriel Sutton was another woman just up the road here and

she won the Queen of the Plough, she is my godmother. It was she who drove my mother to the hospital when she was expecting me. Muriel married a ploughman, Ronnie Sheane, who represented Ireland at the World Championships. He won at national level and he got onto the world team and that was a romance

Liam Hamilton

that would have started through the ploughing.'

Although many ploughmen started when they were only children, others didn't get involved until much, much later. One of those was Liam Hamilton from Ballinagran, a latecomer at fifty-eight years of age. 'I worked on the railways, I worked on

the buildings and I worked in the Wicklow Corn Company, I looked after cattle and racehorses – everything but ploughing.'

Liam's wife died in 2001. Then a friend, Eugene Stephens, invited him to have a go in a Vintage match in Ashford, using Eugene's tractor and plough. '*Begod*, I thought, *I'll have a go at it myself!* I enjoyed it. I didn't win that day!' Liam has now become a successful and enthusiastic ploughman, competing in twenty-five matches a year from Wicklow across to Tipperary and up to Roscommon.

He went on to qualify for the Nationals and has now competed on seven occasions, taking the Vintage title in 2009. He then moved from vintage to conventional ploughing and 'I've been

John Hurley with a friend

winning Wicklow ever since'. In 2016, he took the All-Ireland title in the Junior Conventional Class. Aged seventy-one, Liam says he'll keep at it now. 'I'll have a good go at it over the next couple of years,' he says.

Then there are those who are content to plough the family farm but either don't have the time or the interest in competitive ploughing.

'It was hard work, ploughing wasn't a gentleman's job!' That's eighty-eight-year-old John Hurley from Clonpaddin. Today John is famous for his herd of pedigree Holsteins and some of them are film stars in their own right, having starred in the romantic comedy *Leap Year*, though they are probably best known for appearing in TV adverts for Kerrygold and Budweiser.

Caitríona Murphy in the *Irish Independent* described in June 2011 how the Clonpaddin cows were filmed. 'Having their hooves polished by dedicated farm workers, grazing on pasture while listening to classical music, being sheltered from the rain by umbrellas and being hand-milked beside a warm stove. Animal wrangler and former guitarist with the Boomtown Rats Gerry Cott spent a week training the cows to walk under umbrellas for the popular ad.'

John's father was not a farmer and, in fact, John was born in the town of Arklow in 1929. But when he was seventeen his parents, Henry and Elizabeth, bought a 66-acre derelict farm at Coolmore for £200 – 'And the man who was selling it gave my mother a thrupenny bit for luck and he was so mean that he held it behind his back so he wouldn't see it going!' Then the 90 acres at Clonpaddin came up for sale and the Hurleys bought it for £900.

John became an accomplished ploughman on the farm, 'my two legs were worn out at it'. At one stage, they had six or seven horses and when John was fourteen he used to take them to work with him in the forestry in Glenart Castle. He remembers his mother driving over in the car and he'd be holding the horses out the window as they trotted behind the car. 'You wouldn't be doing that now!' He recalls one particular horse, Paddy, who was an intelligent animal. John was earning 12 shillings and 6 pence a week then.

> He remembers his mother driving over in the car and he'd be holding the horses out the window as they trotted behind the car.

But his pocket money of two shillings had a bit of an accident one Sunday when himself and his friend walked to Barniskey mass up the hills. 'I had two shillings to go to the sports and one pence for the priest. Didn't I make a mistake in mass, I put the two shillings in the priest's box! "Come on back and we'll ask him for it!" said my friend. "Oh begod," I said, "I won't do that!"'

In 1954, John established his first herd of pedigree Friesians and over the years he has been extremely successful, as evidenced by every spare surface in the house being covered with trophies and walls smothered in rosettes and photographs of his top cows, with names like Cradenhill Linjet, Croagh Susie Mist and Clonpaddin Jet Jolly. He has developed the herd through a system of cow 'families' and began importing animals as far back as fifty years ago when he travelled to Canada and paid £5,200 for a calf for his fledgling herd. John still works on the farm alongside two of his nine children, Gary and Pat.

He has had little time to practise competitive ploughing over the years, especially after the death of his father at just sixty-five years of age. But he did find time for off-farm ploughing, competing in local matches, winning a number of awards and even coming first on one occasion. 'But isn't it marvellous?' he says. 'Aren't we a wonderful country! We can go out and meet the world on the ploughing field and hold our own.'

A Day at the Ploughing: Roundwood

As I travelled around the country, people kept telling me that if the book was to be any good at all, I'd have to try my hand at the ploughing! Well, the opportunity came when I went along to the local match in Roundwood in County Wicklow. Throughout the year, there are local competitions held in each county in order for ploughmen to qualify for the National Championships. On a wet and windy day in February, I went along to one of the Wicklow qualifiers.

'Take a run at it, girl! Take a run at it! Back up onto the road and you'll make it!' I had arrived at the Roundwood Ploughing Match where, even early on, the entrance through the farm gate was a sea of mud. I had the boots and the rain gear, which was just as well as the romanticism of the 'soft day' overlooking Roundwood Reservoir turned to a paella of fog, wind and sunshine. The tractors were grand. After all, this is their territory, up and down, up and down, through banks of clay and scrub.

'Ah, Janey, look who's here! Are you still alive? I thought you were dead, I was watching the *Wicklow People*.' That's how horse ploughman Kevin Doran greeted his friends as they arrived.

'It's only wishful thinking,' they replied. 'He's doing the humanist thing, a wicker basket, the whole lot.'

'They'll bury me standing up!' retorted Kevin.

He was keen to introduce me to his horses, Tom and Womble, an odd combination of names. 'I used to have Tom and Jerry, but Jerry retired.'

Telling me I wouldn't find 'any calmer horses', Kevin decided to let me try ploughing for myself. So, boots in the furrow and gripping on the horses' reins, I was encouraged, advised and threatened not to make a mess of the furrow by Kevin and his teammates, Donal O'Keeffe and Martin Austin. There were on-lookers but I hoped they were admiring the horses and not my new-found skills! Meanwhile Tom and Womble showed off – they weren't going to let the side down even if a rookie farmerette was driving them and, truthfully, they were working beautifully, oblivious to the strange hand on the plough.

Getting to know Tom and Womble

It's a wonderful feeling trailing through the soil behind two powerful animals. It made me understand why many city folk have now become weekend ploughmen and women, usually tinkering with old tractors and getting back to their roots at the end of a busy working week.

And as to Tom and Womble's genealogy? 'They were bred in Carnew, trained in Carnew … well, near enough, they live in Carnew!'

Kevin won the Special Horse Plough Class in 2008. 'I was supposed to go to the World Championships,' he says, 'but there was a bit of a mishap with the passports. The young lad here was to come with me but his passport was out of date and I wouldn't go without him.' Really, it is a team sport.

The plough the horses pulled had come through three

generations. It originally belonged to Mick Redmond of Kilmuckridge in County Wexford. He won an All-Ireland with it and then Kevin's grandfather and father ploughed with it, so Kevin said he'd give it a go too. 'I sort of got to like it and I kept going.' Kevin has a small farm in Carnew so the horses are working away all the time and he believes this is why they are so calm and professional that even a nosey reporter could have a go!

Trying out the ploughing

Loy Ploughing

A quiet section of the field in Roundwood was marked out for loy digging. Eugene Stephens from Ashford explained that the loy is an old spade which goes back to the famine that's used for hand ploughing. His own loy was made in 1947.

Different counties had different types of loy – this is a Leitrim loy – but they were quite common in Longford, Sligo, up that country. There were very few in East Wicklow because of our type of soil and probably we could afford a horse plough. In the west, they couldn't. There's even a right-hand loy and a left-hand loy, it depends on which foot you're inclined to dig with. My mother comes from Ballinamore in County Leitrim and, when I was a tiny tot, we used to go up there. There were two old farmers, bachelor brothers by the name of Flanagan and they were famous loy makers and I used to go down to their workshop and watch them. It's one piece of ash the handle is made from and then a small steel tip on the end, about three inches, made by a blacksmith. I got interested watching Francey Flanagan making the loys. Years later, I asked my aunt if he was still making them. He had two left and remembered me as the little garsún from Wicklow. He sent me the loy and the other went to the potato museum in Holywood, County Down.

Jim Heneghan from Bray, County Wicklow, starting his plot

Eugene says he comes about seventh in the Nationals every year – 'I need more practice.'

Another loy digger was Jim Heneghan, originally from County Mayo but now living in Bray, County Wicklow. 'When I retired,' he told me, 'I mosied down to the All-Ireland in Athy. I saw a guy doing a demonstration down there and we had a chat. I told him that in Mayo when I was a garsún, there were old men had the loys and they wouldn't use anything else.'

Later, Jim drew a sketch of a loy and got a carpenter to make it for him; the timber was ash from Laragh and Johnny Gallagher, a retired blacksmith, made the tip.

Eugene Stephens from Ashford, County Wicklow, digs out his plot

The Tractors

From early morning, the tractors clung to the sides of the hill, funnels and cabs breaking proud of the mist, rivulets of rainwater shooting the breeze and running into the freshly dug, clod-scented earth. Smart, shiny new machines and older vintage models that announced their arrival into the field belching grey smoke out over the laneway.

Donal O'Keeffe gave me a running commentary.

We scraped out the width that we need to have to rise our middle. When we have that done, we scrape the two marks up

'In Mayo when I was a garsún,there were old men had the loys and they wouldn't use anything else.'

and down, then we change the settings on the plough and we put a pint on [this is the cutter], that allows it to cut down into the soil and for the sod to turn over. We had various stoppages up along because you're continually adjusting because the ground is up and down a small little bit. Then the third thing we did is called *brushing*, that means we take the point off again, we change the bar on the front of the plough and we get the two wheels to fit in the full width of the furrow where our middle was raised. And then we brush it in, which means you close it up and the term that we mostly use is that a mouse can't run from one end to the other! So that's done, so now we're putting the point back on again ... Are you following me?

Now he's setting up to put his first furrow back up to join the middle and that's when the fun starts!

Robert Mitchell from Carnew, County Wicklow, on his vintage tractor

Watching the fun was Tom Somers, a sprightly looking eighty-year-old from Brackshill in County Wicklow. 'It's over sixty years since I ploughed with a horse and twenty since I ploughed with a tractor,' he said, before confiding, 'I miss the horses. I miss the whole bloody lot, but I got too auld for it and I'm not able to walk as well as I used to.'

Robert Mitchell was competing with a vintage tractor, a standard 1942 Fordson. 'My father acquired it many years ago, it was heading for the scrapyard and he rescued it.' Today, Robert's own collection in Arklow runs to thirty tractors. He won an All-Ireland two years ago.

Another competitor, Michael Shanahan from the Oylegate-Glenbrien Ploughing Association,

Tom Somers follows the plough on a soft day in Roundwood

won three All-Irelands in the Vintage Two Furrow Mounted Class in 2006, 2007 and 2008. He competed with a Massey Ferguson 35 with a Ransome plough made in 1959 'and it's still running and competing'.

His brother Noel competed in the same class in 2004 and, he says, his father competed in the Nationals but 'died before he could reach his full potential'. However, Michael's Uncle Joe took the All-Ireland Senior Class in 1976, and represented Ireland on six occasions at world level. He said it's hard to explain the attraction of the plough but 'it runs in the veins like the blood

'I miss the horses. I miss the whole bloody lot.'

when you remember being a small boy with your father and your uncles, it's just something that's there, it's handed down from father to son'. He recalled the success of several other club members – John Somers who was second in the world championships and who won several All-Irelands, Michael Scallan, Jimmy and Joe Shanahan, world champion John Whelan and Martin Kehoe, 'the Michael Schumacher of ploughing'.

There was a team of four judges who wandered discreetly around the field, watching everything and huddling into little conclaves every so often. I introduced myself and asked them to give me the five-minute guide to being a judge. They explained the importance of the opening split where they look for straightness and evenness 'and enough room for the middle to fall back into.

Eugene Buttle from Blackwater, County Wexford; Gerard Walsh and Murt O'Sullivan from Kilmuckridge County Wexford; and Harry Williams from Enniskerry, County Wicklow

The middle goes into it then, a three sod or a four sod middle.'

The judges in Roundwood were Eugene Buttle from Blackwater, County Wexford, Gerard Walsh and Murt O'Sullivan from Kilmuckridge and Harry Williams from Enniskerry, all of whom have a lifetime of experience under their belts. They were happy with the standard and delighted to be meeting friends they hadn't seen for a year.

'It runs in the veins like the blood when you remember being a small boy with your father and your uncles, it's just something that's there, it's handed down from father to son.'

Murt told me his own story. He was at the Ploughing in Enniskerry in 1965 when he was fifteen years old. They stopped for a bite to eat at Toss Byrne's pub in Inch on the way home. 'This man was sitting beside me at the counter and he asked me, "What's a good job of ploughing?" I told him, "To have no grass showing and a good, even job, really it should be like a pair of corduroy trousers!"' His own father, Mike, kept horses and was 'a brilliant horseman', remembers his son. 'Kind. But if you meddled with the horses you were a dead man.' Mike was also the coach for Ned Jones when he won the first All-Ireland.

The evening was getting darker, the mist was moving in and still they were ploughing. 'We may go for the flashlamps,' said a judge. And the others remembered finishing at a match in Laois one day. 'It was a wild bad evening, boiling wet. We had about five plots left and the boys nearly ended up in the dark. We were just finished in time.'

Darkness started creeping up from the depths of the lake beneath our feet, the drizzle turned to rain, the tea stall packed up and the judges were getting worried. 'Darkness! We may go for the flashlamps.'

The horses that lead double lives

All winter they work on the farm, pulling ploughs and competing in matches around the country. Then comes summer and, like excited schoolchildren released from the classroom, they cast off the shackles of the plough to trot along country roads pulling gypsy caravans for the delight of tourists. Their strength and lovability make them great ambassadors for the country as they pause to see the waves crashing along the shore, nibble grass in strange fields and pose for tourists' sentimental selfies. Yes, the plough horses are off on their summer holidays.

Some farmers do have their own teams of horses but others borrow them for the winter months from horse-drawn caravan operators, such as Clissmanns in County Wicklow. Neasa Clissmann, who now runs the business with her father, Dieter, told me how it all came about. 'My parents set it up in 1969, my dad actually used to work for Fáilte Ireland in Frankfurt and he decided he wanted to move home, so they came back and set up the horse-drawn caravans because other people were doing it in Ireland at the time. He was familiar with promoting Ireland because of Fáilte Ireland. He had no horse experience but just came back and bought the farm and jumped right in! He employed people who had experience and the expertise that he was lacking. That's how it all came about! My mum Mary passed away in 2011, so I've been doing it since then. So I guess it's kind of second generation.'

And that's where it gets interesting because, back in the sixties, things were very simple. 'People would take a caravan and a horse and head off down the road and just ask a farmer if they could camp up overnight. These days, it's a lot more structured, we've farms where they pull in at night and there's showers and toilets. In the sixties, it was generally a dunk in the river and you were washed!'

Neasa is the second youngest of six children and, during school and college holidays, they all worked with the caravans. She is now the only one left and has taken over from her parents. Dieter is in his late seventies and, according to his daughter, he's there 'as a wise old head'! 'My dad set up a system where we would use the horses all summer ourselves, and he would lend them out to people free of charge for the winter, as long as they were well looked after. Some of them were used for forestry work, local men would come and borrow a horse for the winter and it kept the horses fit. Sometimes, we'd loan a horse to be ridden. It's a good system, you get to have your own horse for six months without actually having to buy it. And that's where the ploughing came in. There were people involved in hobby ploughing. Their farms would be sheep or cattle, dairy or whatever and they'd just have a love of working with the horses, so they'd borrow one for the winter. I remember, as a child, there would be fifty horses on the farm in the summer, and there could only be ten left in the winter! They were all loaned out, and very well looked after. It was great for

us, because the nature of the business is that our horses are handled by different people every week, and some horses can get in the habit of being handled by the one person and getting reliant on that person, whereas by lending them out to different people they were getting more experience. It's a bit like a baby being passed around! They're kind of comfortable with it.'

Neasa talks about the 'subculture' that surrounds the horses and the ploughmen. 'I love it because you're dealing with all these cosmopolitans from Berlin and Paris and Frankfurt, and then, on the other side, you're dealing with the ploughmen and only for the job that I'm in, I would never come across them. I would never meet them. And they're so lovely, and their hearts are so in the right place, and it's a pleasure to know that people like that still exist. We might only see each other once a year but there's a warm, reciprocal thing there. It's a nice part of Ireland and to know it still exists is lovely.'

Lending the horses out was an interesting move and the Clissmanns were alerted to it after Mary bought fifteen horses from another horse-caravan operator in Kerry called Slattery, who had retired. 'Mum decided she was going to buy all the horses because we were finding it harder to get really well-trained, quiet horses. We sent a friend down with the horse lorry, and in two deliveries he brought all the horses back. And then it came to the end of the season and Mum was getting these phone calls from these lovely guys down in Kerry who she had never met, and they were going, "You have my horse!" And she was like, "No, no, I bought them fair and square!" Then she discovered that these guys had been in the habit of getting the same horse every year, in the same scenario, with the Slatterys. And, sure, they were kind of bound to these horses and were very fond of them and they were wondering if there was any way that they could have the same arrangement, so that's how that came about! Once a year, the Kerrymen would share a lorry, there would be a few of them – J.J. Delaney and Denis Kelleher and a few others – and they'd pay a man to come up with the lorry and take them in one go and then

'We might only see each other once a year but there's a warm, reciprocal thing there. It's a nice part of Ireland and to know it still exists is lovely.'

we'd get a call in the spring that they'd be on their way back up. So that's how it came about, their double lives.'

Another revelation for the Clissmanns was that the ploughmen had different names for the horses. 'We have one horse called Tango and I was ringing one of the lads and saying, "Are you bringing Tango back up?" And he says, "Which one is Tango?" So I said, "The black one", and he said, "He's called Lazy Simon!" So they literally have a different identity, two different lives.'

J.J. Delaney still ploughs at the Nationals with two of Neasa's horses (he won the Horse Class in 1995 in Ballacolla, County Laois) and Neasa now goes along to watch Toby and Paddy, 'like a proud mammy'.

'When a horse reaches the end of its working life, they retire with us, we don't sell them on or anything, they stay on the farm. One or two of them we lend out to a petting farm. I say it's like horse heaven because they just get endless carrots and apples from the little kids, it's so lovely! So typically they retire with us, but the Cork horses, as I call them, if the lads want to keep them down in Cork, I let them. Sometimes, the horse can do a bit of ploughing with them, it might be too old to pull our caravan and they'll stay with the lads until they pass away, and I know they're in really good hands and they adore them. I'll always get updates, like "Lazy Simon is doing great, he's not doing as much as he used to but he's still in good nick."'

Neasa says she realises that she is blessed to live in this snatched spot in time in rural Ireland. 'We live in the middle of nowhere, but the world comes to us. That's how I always put it. We get the countryside life without being isolated. Its cosmopolitan countryside! It keeps it interesting.'

Donegal

There's a whiff of romance about the ploughing in Donegal. I'm talking to Alan Simms, one of the legendary Simms family, and he's telling me how he met his wife-to-be, Tara Jacob, in 2011.

Alan was getting ready for a ploughing match in Donegal when he got the job of putting together a first aid kit. He went along to the local pharmacy where he met Tara, who was the local pharmacist, and they were married a year and a half later. Not to be outdone, a friend of his, Peter Buchanan, was competing in a match in north Donegal when he fell in love with Patricia McHale, a spectator from Mayo, and the couple are now married with two children. Clearly, Alan sees ploughing in Donegal as something akin to the matchmaking in Lisdoonvarna when he tells me, 'We have to keep reminding the ladies that sometimes ploughing brings benefits for them too ... and we need their support!'

> There's a whiff of romance about the ploughing in Donegal.

But back to the actual ploughing. Alan is Secretary of Donegal Ploughing Association and Treasurer of the Lennon Ploughing Club. His father Norman, who came from a mixed

Alan Simms meeting Michael D. and Sabina Higgins before winning the European Classic Ploughing Championships

farm in Milford, was involved in setting up his local club. Alan was one of ten siblings, seven brothers and three sisters. In recent years, five of the Simms brothers have been collecting titles both at local and national level and frequently appearing in the top five at events in Conventional, Reversible and Vintage classes.

Norman Simms on the family farm

In 1966, the NPA introduced awards for provincial winners from the Under-21 Class and the Simms have made it their own. Gary started off, winning on three occasions in 1997, 1998 and 1999, and Matthew came next, winning in 2004 and then going on to take the Under-28 Conventional Class in 2007 and 2008. Lee, the youngest brother, won the provincial award in 2013, 2014 and 2015. Meanwhile, Alan took second place in the Vintage Class in 2016 and was twice the winner of the European Classic, which was hosted in Northern Ireland in 2011 and then in the Republic in 2014. Andrew has already competed at national level and Mark is described as 'still upcoming' but hasn't made it to the Nationals as yet. 'We have youth on our side,' says Alan. 'The youngest is twenty and Gary, the oldest, is still only forty, so we cover a twenty-year span and we're still learning every year!'

Alan says his father was 'happy enough' with their results. He had taught them all how to plough and made sure that they didn't come head to head against one another in the local matches in order to give each of them the best chance to qualify for the Nationals. 'He would have been disappointed if two of us were trying to compete for the same place!'

The boys' father took a break from ploughing competitively 'but he's back now in the Vintage Class – he actually qualified for the Nationals, but chose not to compete! He's letting his six sons head off.'

Alan says Donegal is regarded as one of the strongest counties, well respected by the NPA for the level of achievement and dedication that goes into the ploughing. 'We have a few underage titles,' he says, 'but we're waiting for the seniors.'

And although the Simms still plough on the family farm and some of them are agricultural contractors, Alan says for most competitors these days, ploughing is a hobby. 'Just another sport. Some people have athletics, we have gravitated towards ploughing. But you cannot overestimate the power it has to bring communities together.'

'You cannot overestimate the power it has to bring communities together.'

We have four matches within the county and we regularly travel to neighbouring counties, including in the North. We all know one another, Donegal clubs support their clubs and vice versa. The judges and stewards help their counterparts and we invite them down to judge with us.

Competitive ploughing will survive if it has the right supports behind it. The current system of ploughing for tillage has changed and will continue to change. Quantity has taken over from quality in relation to getting things done quickly and efficiently. But the Ploughing is a good release for the farming community and it's all bound together by what the NPA has achieved in the past eighty years.

The six Simms brothers

Loy digging

One area in which Donegal has swept the boards is loy digging, the oldest method of all. Since 2004, Donegal competitors have taken twenty-two titles in Senior, Junior, Under-21 and Ladies' Loy Digging classes. In 2015, they had an amazing result when all four events were won by Donegal – Brendan Byrne took the Senior Class, Kieran McDaid the Under-25, Evan McGirr the Under-21 and Kathleen Donaghey the Ladies' event.

Many of the winners have several titles to their name and one of them, Marian Boyce, won the Ladies' event on four occasions in 2008, 2009, 2010 and 2012. She grew up on a farm and has ploughed with a tractor and even made it to the Nationals Farmerette Class in 1993. She only took up loy digging in 2006, when she watched a demonstration at a ploughing match. 'Gerry Mallon from Kilmacrennan was showing people how to use a loy and he gave me a go of it. I liked it and I just took it up from there. It's a heavy implement but I really enjoyed getting

out and about in the fresh air with my fellow diggers, having the craic and a bit of camaraderie together. The reason Donegal has been so successful is because we're just interested and enthusiastic and we enjoy it. I only practise if I make it to the Nationals. A couple of weekends beforehand, I'd go out and practise just to get the strength back in my arms. There's a great atmosphere among the loy diggers, we help one another out, there's a couple of wee girls involved now and we try to grow the interest. Some people remember their fathers using a loy. It was used a lot in Donegal in the old days, to till bad soil and rough ground.'

It's also a very cheap way of getting involved in ploughing. Instead of spending many thousands on a tractor and a conventional plough, 'It's just a matter of throwing the spade into the car and off you go.'

Next, I spoke to Gerry Mallon, the man who inspired Marian to have a go and the person who actually makes the traditional loy spades. He started his ploughing career in two-furrow-style on a tractor. 'But in those days, twenty years ago, I wasn't about home and there was only one tractor and the brother was ploughing with it. I was away all the time working in Dublin.' Then Gerry Hunter went to a demonstration of loy digging in Donegal given by two Sligo brothers, Alex and George. 'They asked for someone who might be interested in taking it up and I volunteered.'

Marian Boyce

Gerry then made his first loy in a bedsit in Dublin from a template given to him by Alex Hunter. He went on to develop the design and, to date, he has made over a hundred loys, though he can't remember exactly how many. When I talk to him he has just delivered two loys to someone in Carrick-on-Shannon to be hung in a pub in London. 'I'm almost the last loy maker in the world. They are made from Scandinavian ash and I'm blessed that we have a saw mill in Kilmacrennan and Paul McFadden is very generous in getting me the best ash you could ever ask for. A neighbouring man shapes the wood for me and I craft it from there. The metal part is made by a very special blacksmith in, believe it or not, a fully laid-out working forge in Leeson Street in Dublin. This man worked all his life making railings for Dublin city houses.'

Gerry Mallon

After Gerry got the bug, he started loy ploughing competitively and now has four titles – two junior and two senior – under his belt. He had already won at senior level when he managed to break his loy in the middle of a competition and so was demoted back to junior level! He loves making the loys and says any chance he got he was making them and he got great satisfaction out of seeing up to twenty-two young people, aged eight to twelve, digging away with his loys at a training session in Donegal.

Each loy is individually crafted. A right-handed digger digs with the left foot and a left-handed digger digs with the right foot. But it's not an exact science, there have been exceptions. I measure a person's size while I'm talking to them. I stand in front of them and measure down to their chin the number of buttons on my shirt. I put my arm straight and get them to put their arm beside mine and I figure out the measurements, a fist below an even fist or a fist above. I then go and make a loy to suit that person.

'I measure a person's size while I'm talking to them.'

Gerry lives in Dublin, where he works on a construction site. He is a carpenter by trade but says, 'You have to turn your hand to everything on a building site.' He would like to get back to living in Donegal eventually. 'I wouldn't like to become a spider on the wall in Dublin. I would have gone to London years ago, but what kept me here was my input to loy digging, the coaching and the competitive side.'

He says he has beaten most diggers in the country at some stage or another 'but every dog has his day and the people I coached – I now have terrible bother to beat them'.

Gerry still gives demonstrations at vintage days around the country, and in 2016 he decided to take his demonstration to the World Championships in York in England, given that loy digging is *not* a class at the world event. He just went into a field and started digging. A crowd gathered. But he had chosen the wrong field and was digging away where the tractors were about to start ploughing. 'I was digging where I shouldn't be digging! It's an Irish thing!'

Offaly

This is the first time I've heard of 'Rock On' Paddy. His real name is Paddy Keating and he is probably the most colourful member of the Offaly County Ploughing Committee. His calling is to entertain the crowds at matches the length and breadth of Offaly with his one-man band. He has even appeared at the Nationals in Screggan and, according to County Secretary Trina Connolly, 'He is the life and soul of our presentation nights.'

Trina herself says she joined the same committee by pure accident. Her father had represented Offaly a number of times at the Nationals and took the Three Furrow Conventional title in 2003. 'We were always interested in going to matches when we were kids. Then, one night, there was a meeting and Dad brought us along to make up the numbers. It turned out to be the AGM, and I got voted in as the PRO!'

Then, she discovered that the PRO was also the assistant secretary. Trina has spent seven years as County Secretary and the officers were kept on for an extra two years because of the Nationals coming to Offaly in 2016 and 2017. 'We are an ideal venue for the Nationals, the road network is brilliant, it's a great place for it.'

Offaly is doing very well in ploughing at the moment, she tells me. 'We have fourteen competitors going to the Nationals in 2017, we're a very strong county. Ploughing is a sport on its own, for many people that's all they would do, not Gaelic or anything else!'

Michael Mahon has been representing Offaly on the National Executive of the NPA for twenty-seven years. He comes from Blacklion, at the Blue Ball between Tullamore and Kilcormac, where he grew up on a mixed farm, the eldest of the eleven

Michael Mahon (centre) with William King from Derry and John Tracey from Carlow judging at the National Ploughing Championships, 2012

children of Sean Mahon. He works the same farm today and he and his family have become serious competitors since he came second in the Junior Conventional Class in 1985, in Kilkea, County Kildare, second to his fellow county man Frank Gowing. Fitting perhaps that Michael went on to become director for Offaly in 1989 on the death of Frank Gowing. He was also the National Chairman of the NPA. Normally, that's a three-year stint. 'But the Worlds came when I had the three years done and the executive decided to leave me in the chair for them. Then the Nationals came to Offaly in 2007 and they left me in the chair again!'

Michael says he remembers the first time the Nationals came to

County Offaly has hosted the Nationals six times, including at Screggan in 2016 and 2017

County Offaly, in Edenderry in 1982, then it came to Charlefield, Tullamore in 1987 and then it came to Quinns above in Rath, near Birr in 1997 and then we went back and we came to Annaharvey in 2007 and we're in Screggan, Tullamore in 2016 and 2017. I can't say if it'll be there for a third year, there's six or seven counties looking for it. There's Carlow and Kilkenny and there's Kildare ... we have to run this year first and then when that's done, the executive decides where to go after that. We're looking for sites as it is, but it's not easy to get a site. There was a time when if you could get a big field, you'd run it; now you want 700 acres. But if things go well in Offaly this year, sure, you never know. We could be in the ring for another year!'

I'll tell you now what makes the ploughing so successful. If you ring the NPA at eleven o'clock at night, there will always be someone to answer the phone. Anna May will answer that phone no matter what happens and that is the difference. It came from a very low ebb to what it is now. They're doing a marvellous job, but it's a marvellous organisation led by Anna May herself. There will be big changes when Anna May retires. She has a daughter well fit to run the show and then there's an executive director from each county, and they only have the one vote so no one person can rule the roost. If everyone brings what the club or the county want and it's not a mé féin job, it will survive!

'If everyone brings what the club or the county want and it's not a mé féin job, it will survive!'

Michael's sons Justin and Brian have been ploughing for the past seven years, and Justin won in the Under-21s at national level and Brian has competed at European level, winning the Five Nations on two occasions.

Michael has also moved into judging and has been a judge at the last six World Championships and Brian has been a judge for the past three years. 'When you get to judge, your job is to get all the marks you can for the competitor. That's what you're there for!' I ask him about disgruntled ploughmen and complaints. This elicits a wise chuckle. 'The lad that wins never complains, it's the lad who thinks he's won that does!'

Kerry

'To get good poitín now is like trying to get good horses.'

It's a cold, wet evening in Kerry but I'm in the snug of Benners Hotel in Tralee with Tom Healy, the Chairman of the Kerry Ploughing Association, and PRO Tom O'Mahony. We're drinking coffee and they can't resist the story of the man who transported Kerry horses to the All-Ireland every year. He always collected the horses near a local pub. 'One night he was a small bit late collecting them, and all he could see were the two stables, two horses and two pairs of wellingtons,' says Tom. 'He knew where the lads were, inside in the bar, so he decided to get started and picked up the wellingtons to fling into the lorry. "Jaysus," he said, "the wellingtons were awful heavy." He had a look into them and there was a bottle of poitín inside each boot! But they were West Cork lads.'

We travel back to 1856, when ploughing started in Kerry. Lord Crosbie and Lord Listowel, two local landlords, held competitions among their tenant farmers to see who was the best ploughman. The prize would be a harrow or a spade or a shovel 'or something to encourage the tenant farmers to be productive and that they would be able to grow stuff and able to pay their taxes'. Then the records disappear until 1911. That, they say, was because of the foundation of the Land League by Michael Davitt in 1879 when tenant farmers were paying unjust taxes to the landlords. From 1911, the ploughing matches started again but stopped in 1918 with the Black and Tans and the Civil War. There are no more records until 1932 but a match has been run every year since to the present day.

I ask why a place like Abbeydorney has produced so many champions. Tom O'Mahony answers.

Well, the Nationals started in 1931 and Kerry competed in

Another of Abbeydorney's champions, J.P. Shanahan, who won the Senior Conventional title in 1951

1932, but it wasn't until 1941 that the first All-Ireland title was brought to Kerry and the reason for that was that the two Leinster founders, Denis Allen and J.J. Bergin, had a Leinster style of ploughing that was totally different to the Munster style. The Munster style was a shallow flat type of ploughing whereas the Leinster style was a deeper and rougher style. The judges at the time, of course, were Leinster men, right, and Kerry and Munster had to adopt the Leinster style in order to win. So, undaunted, the Kerry ploughmen brought down a man called Redmond from Wexford to give them a few pointers and they took off then and won in 1941! To date they have won forty-three titles at national level by twenty-five individuals – 2015 was a bad year with no All-Ireland titles but we were always in the first four positions. You know yourself, there's an awful lot of luck in this game, first of all by the plot that you're drawn. The plots are drawn by the NPA, so it's the luck of the draw. There's no cheating but maybe a bit of favouritism by the judges – but we have Kerry judges now!

'The Munster style was a shallow flat type of ploughing whereas the Leinster style was a deeper and rougher style.'

Tom O'Mahony is a nephew of the legendary horse ploughman Pat O'Mahony who won three All-Irelands in 1953, 1960 and 1963, and he taught his nephew to plough. 'Also,' Tom says, 'the salesmen with the tractor companies would normally give you a bit of coaching.'

The name of Martin Slattery is mentioned. He was a salesman with Benner's Garage in Tralee and he seems to have introduced many of the Kerry ploughmen, including the farmerettes, to their tractors.

The men tell me the reason there is so much interest in ploughing in the county is because there is so much tillage, especially in north Kerry. There are currently five ploughing associations in Abbeydorney, Ardfert, Ballyheigue, Ballyduff and Causeway. Back in the fifties, there was also a ploughing society in Killarney and men from Killarney took six All-Ireland titles. It has been resurrected again as a Ploughing and Vintage Festival and, hopefully, in time it might become affiliated again. But there's no getting away from the fact that, in 1954, Killarney hosted the first World Championships to be held in Ireland. It was only the second competition – the first had been held in 1953 in Ontario, Canada. 'I don't know how we got it,' admits Tom Healy, 'obviously there was someone with a bit of pull!'

The top ploughman at the 1954 World Championships was Hugh Barr from Northern Ireland with Leslie Dixon from the UK the runner-up. Hugh Barr went on to keep his title in Sweden the following year and in Great Britain in 1956. The winner of the first World Championships, held in 1953 in Canada, was Canadian James Eccles. He competed in Killarney in 1954 without success. At that time, Kerrymen John Joe Egan and Pat O'Mahony were two of the top horsemen in Ireland. Tom Healy says they held that position for several years and 'they were as well known then as any member of the Kerry football team'.

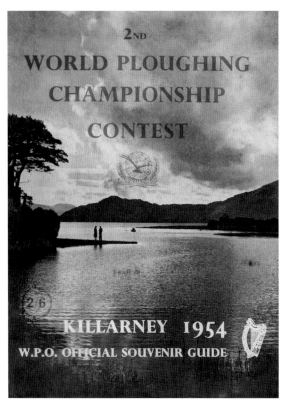

The National Championships have gone to Killarney on two occasions, in 1939 and in 1961. On that second occasion, Noel O'Reilly reported in the *Irish Farmers Journal*: Killarney itself proved

*John Joe Egan competing in the Senior Horse Class
at the Nationals, 1961*

an ideal venue. Despite the heavy rain on the previous day, the plots were dry and turned up in good condition. However, the surface of some of the plots was very uneven – particularly in the Two Furrow Senior and Junior sections, providing competitors ample room for exercising their skills.

*Pat O'Mahony, All-Ireland Senior Horse
Ploughing champion*

And O'Reilly praised Eileen Brennan of Laois, the winner of the Farmerette Class.

> Her ploughing has much improved since last year, and the verdict was well merited, although closely challenged by Angela Galgey of Waterford, who was only two marks behind … For the first time in the past six years, the stewarding ran very smoothly, and the spectators did not encroach on the plots. Numbering the plots in the catalogue also proved very handy in recognising the plots, but many felt that the old way of putting a number on the competitors' backs should also be retained.

The paper's photographer, Ambrose O'Mullane, had a photo in the same edition of Mrs B. Grosvenor, Chairman of the local organising committee, setting off a flare to start the championships.

'What about the social side and the romance?' I ask the Kerry ploughmen.

'Sure you'd put your eye on a very nice girl, but sure she mightn't look at you at all!' says Tom Healy. 'The Queen of the Plough dance was always a great occasion and there were probably romances out of it! There were always dances on the first and second nights of the ploughing and they were very well attended. But that's gone now. But the ploughing is still a great place to meet people from every county in Ireland. The city people love coming to see the horses … and the local radio is great now, it's a blessing because if you had to postpone the match due to bad weather, you can contact the people straight away with the local radio.'

'But the ploughing is still a great place to meet people from every county in Ireland.'

Next, Tom O'Mahony produces a little box from his pocket, swings it open and asks me to guess the contents. It turns out to be a solid-silver medal presented by the

Royal Dublin Society for the best turned out pair of Irish draught horses in 1939. It was made by Weirs of Dublin and is a beautiful thing. Tom tells me that it was presented to a man from Abbeydorney. But there is an amazing story behind the medal. 'There was a house being sold and they were cleaning out the house and emptying the contents into a skip. Someone spotted the little red box inside the skip and they opened it and saw it was to do with the Ploughing and they brought it to me.'

The silver medal

Our discussion turns to the future of the Ploughing and whether or not it can survive.

'That's a hard one,' says Tom Healy, 'first of all tractors have got so big that it's hard enough to get a suitable tractor small enough for ploughing, and the ploughs are all big ploughs. For a young fella just starting ploughing, it's a pretty expensive game whereas, when I started ploughing myself, we just had the plough we were using at home and the tractor, there was no such thing as even changing the wheels, it was a small tractor anyhow. There's nothing like that now, you must have this competition plough, you have to have a special plough set aside for the ploughing matches and even a special tractor set aside, they'd never take the plough off the tractor. It's hard enough for a teenager to persuade their dad to fork out maybe £15,000 or £16,000 just for equipment.'

'Good horses are also hard to come by – Irish draught or Clydesdales,' says Tom O'Mahony. 'The main attraction of the match is the horses and, long ago, every farmer had horses and they were out ploughing and working the fields every day, and the man could actually stand behind the plough and the horses would walk away because they were so used to it. A ploughman now gets two horses at the competition and they mightn't have pulled anything for the past nine months! Compare that with the All-Ireland here in 1939 when there were fifty-two pairs of horses.'

On that occasion there was a shortage of horses in Killarney and they had to bring them from north Kerry by train.

'They were carried in freight carriages, a special carriage they had for horses, cattle and pigs. The old horses that time were quiet because they were working every day whereas if you got a horse now today and put him into a train ...'

'But if the horses did die out,' says Healy, 'it wouldn't be the same, that'd be a right downer altogether.'

In 2000, the people of Abbeydorney completed a millennium project to remember all the men and women of the parish who had taken part in the ploughing over the years. It was the leading parish for ploughmen, with a total of seven All-Ireland title holders. I went along to visit the memorial with some of the locals. It has a Pierce horse plough mounted on a stone and three tablets with the winners' names. They tell me that the ploughing was always a huge celebration in Abbeydorney, the entire village got involved.

Tom O'Mahony fills me in further. 'There was a lot of celebrating in the old days. The women were at home looking after the farm while the men were ploughing, but they took fierce pride in the horses and they'd always be polishing the harness. If they

Street celebrations in Abbeydorney to welcome home the town's champions from the Ploughing, 1954

were close by, they would arrive at the field and bring the tea and sandwiches. Sometimes, the men would go for a few drinks and end up sleeping in the hayshed for the night.'

He remembers the night before the ploughing match in Abbeydorney in the late fifties. 'All the society members would come together in one house and they'd have so many pan loaves, they'd butter all the bread and make different sandwiches, mainly ham and salad. The following day, we'd all go to the match and there was a horse and cart and an old creamery tank full of tea and they'd give out mugs of tea and sandwiches to the people. There'd be a tractor trailer at the match too, it'd have a canvas over it and they'd be selling bottles of beer and whiskey.'

It's hard to say goodbye to Abbeydorney and I stand for a moment at the memorial and read the names of the ploughmen, some long gone, who brought so much honour, so much celebration to this small place. They'll certainly never be forgotten as each generation continues their story.

Moss Trant

Moss Trant has two Percheron horses and, allegedly, they speak French! He knows, he's tried it, but he says it's really the tone of your voice they understand, the special nuances between a ploughman and his team.

Moss grew up on a farm on the Mile Height in Tralee. His father always had horses but suffered bad health through arthritis, so Moss's last memory of him walking was when he was five years old. As a family, they started working early in life, and he remembers labouring with the horses to cut hay and weeds from about eight years of age. 'The last time we cut the hay with horses was in 1969. We had one horse ourselves, Dolly, and we used to get a loan of another from a neighbour.'

In 1972, they switched from the horse to a tractor but, strangely enough, Moss moved back into horses in the mid-nineties. 'There was an old man, Mike Dineen, back in

Moss Trant ploughing at the Nationals where he came third place in the Senior Horse Class, 2014

*A new Guinness World Record was set at the 100th Annual
Mullahead Ploughing Match, 2015*

Causeway. I used to take the horses to the ploughing matches
for him and he was on to me every year to know would I take up
the competition. One year, I got a loan of two different horses,
borrowed off the jarveys [the horsemen who give horse and
trap rides to tourists around Killarney, Moll's Gap and other
scenic areas of County Kerry]. They wouldn't be using them in
the winter. But we couldn't get the pair to work properly. So I
brought in two horses from France, Percherons from Normandy.
A friend of mine, Tim Sampson in the UK, had Percherons and
I went over to him on several occasions and I liked the animals.
They're a very good animal to work. But they are rare in Ireland,
I was only the second person to bring them in.'

The animals came with their French names – Genece and
Gitane. In France, horses are named for the year they are born,
each year gets a letter and horses are named accordingly. Moss
worked the horses and bred them. 'Two mares, both of them in
foal. Mares are a bit temperamental, but they're better when
they're pregnant, way tamer. What I have now is Percheron, but
geldings.'

By the time the alphabet had got to *L*, Moss had a foal called Laurence bred in 1998 who was struck and killed by lightning.

Moss started entering the local competitions in Abbeydorney but, he says, 'There was no competition, so I came in first for five or six years! I was regularly among the winners.' Moss has now qualified for the Nationals on five occasions and third has been his best position so far – that was in 2014. He says he likes a challenge and he was involved in a Guinness Book of Records feat of getting forty-six teams of horses to plough together. This was in 2015 on the Richardson Estate in Portadown in Northern Ireland. He is also involved in a Charity Hay Making Day for the local hospice, 'making old fashioned cocks of hay, going back to our youth'.

He talks about the great bond between man and horse.

'If an animal trusts you, they're with you, they'll do anything for you.'

It's nice and quiet, there's no sound, only the animal moving and it's a pleasure to work with them, and they know that. They respond to your voice, they just want to know what to do. If an animal trusts you, they're with you, they'll do anything for you. They're friends. But they have their own ways and they'll know if you're not in command.

My horses mean as much to me as my wife and children, they're part of the family as such. When they die 'tis sad. You'd grieve a horse, if you had this animal with you for up to fifteen years and seeing him every day, it's tough.

This love of the horse, this bond with a companion, is talked about all over the country. Even those who have moved on to plough with tractors are still nostalgic for the old days and old men remember the names of their first team, their characteristics, their mood, their temperament.

Westmeath

P.J. Lynam grew up on the family farm in Kilbeggan. He had always worked on the farm with his father Malachy, ploughing with the horses in the years before they got a tractor.

> I had gone to college in England and was there a long time, and, when I arrived home, I got myself a job and a car. Cars were very scarce in the late sixties and early seventies and there was a guy used to call to the house to see us, Michael Grehan. He called in one day and asked my mother if I would drive him to a ploughing match. At that time you did everything your mother told you, this was 1974! I duly drove Michael to the ploughing match outside Tullamore and we walked around all day. After they announced the results, they said there was another match in Westmeath on 14 February. Going home in the car, we were chatting about it and I happened to ask Michael, 'Who's running the match in Westmeath?'
>
> '*You* are!' he says – and the rest is history!
>
> I formed a committee and we ran the match. One of my sisters, Eileen Robbins now, competed as a farmerette.

P.J. had caught the bug!

Then, in 1984, the Vintage Tractor Class started at the Nationals in Ardfert in County Kerry, for which the tractors have to be pre-1950. 'I saw a tractor I liked and it was the same type of tractor I had learned how to drive back on my grandfather's farm in the fifties. I bought it and brought it home and a good friend and myself, we put it together and did it up and I still have it.'

In the beginning, P.J. only practised around the farm. 'I would be a perfectionist, I wouldn't go out in public until I

'Everybody gravitated to our house, a rambling house, people would just call in.'

knew exactly how to do it.' P.J.'s father was ill from the age of fifty, so P.J. ran the farm together with his brother and his mother.

P.J. has very fond memories of his mother, who only died three years ago, at the age of 93. 'After minding us all her life and milking cows, proof that hard work never killed anyone.' He shares a lovely story about his mother's first visit to Dublin in 1995, when he brought her up to see his dad in hospital. 'It was her first time in Dublin and it was Christmas-time, so I brought her to O'Connell Street to show her the lights. She had never seen anything like it before. She was very much a home-type of person and a great conversationalist, everybody gravitated to our house, a rambling house, people would just call in.'

In 1985, all his practice paid off and P.J. qualified for the Nationals in Kilkea Castle, County Kildare, which he won in the Vintage Single Furrow Mounted Plough Class. He went on to take nine more titles – in 1985, 1986, 1987, 1988, 1991, 1992, 1993, 1995, 2004 and 2005 – but then he got very involved in the administrative side of the ploughing with the NPA. The last time he ploughed competitively was in 2006 in Carlow. 'I spent one hour

P.J. Lynam winning at Ballacolla, 2000

away from the office,' he says. 'I jumped on the tractor, dropped the plough and then immediately went back in to the office. I realised then that I had to choose between administration and ploughing, so I figured I'd had enough of ploughing at that stage.'

P.J. has just completed his three-year term of office as Chairman of the NPA. He is an extremely energetic person and makes a very early start each morning. And in the real world, when he's not organising the Ploughing Championships, P.J. is a livestock agent who has experienced the demise of farming first hand.

> I'd be away down the west buying cattle and sheep from people, and the people I had as customers down the years, they're just gone. They come into the mart and they'd be talking and saying, 'I've nothing for you this year, I'm too old and I'm not able and I've nobody at home to help me.' That's the way farming has gone. And if that happens it has to have an effect on the Ploughing at some stage, because the same amount of machinery won't be sold, and if people are not selling machinery they won't want to exhibit because of the costs. There is no answer to this question, there'll always be enthusiasm for competition, so there will always be competition ploughing for a long number of years. But whether the trade area can maintain what it's doing at the minute, nobody has the answer for that.

P.J. works very hard to encourage young people to get involved in ploughing, getting novice classes going and making the machinery available to them to try and get them interested.

I asked him if he thought horse ploughing was dying out.

'Horses dying out? I'm listening to that for the past forty

'Horses dying out? I'm listening to that for the past forty years!'

years!' he says. 'On the second day last year, we had sixteen sets of horses there. But we don't sit on our laurels, we're always working on things and we're promoting the horse ploughing this year with an extra bit of a grant.'

He reiterates that the NPA makes its money from the trade area and the admissions, they don't get any grants, 'We are totally self-sufficient, we owe nothing to nobody.'

P.J. has just returned from a site meeting, even though the Ploughing is eight months away. 'Work starts early on the site,' he says, 'and there's heaps of work to be done even at that early stage.'

But on a broader canvas, P.J. is well aware of the challenges facing agriculture at the moment.

'The future will all depend on what effect Brexit has and what way the country is going.'

With Brexit on one side and Trump on the other, where agriculture is in this country is a very difficult question. Brexit could have an effect on farming – if it has a major effect on farming, people could migrate away from it as they have been doing down the years. The building industry and the boom we had took a lot away from farming. The way you make a living in farming is not great and it's a seven-days-a-week job, 365 days a year, so the young people when the work became available on buildings, driving machines and things like that, they all left, they really all left. When the recession came a few of them started coming back and returned to the farms at home. So, the future ... it will all depend on what effect Brexit has and what way the country is going, if the country is booming, the people will move away from farming.

P.J. is both a realist and an optimist who firmly believes in the future of farming as an industry and, of course, the Ploughing,

*With Anna May McHugh and Eoin Buttle at the end of
Day 2 of the Championships, 2016*

which he says is at the heart of rural Ireland. 'It's the place
everybody wants to be in September. The harvest is finished and
it's a kind of a holiday for the farmers.'

Louth

There wasn't enough work on the farm for everyone, so in
October 1959 at the age of sixteen, Gerry King emigrated to
England. He was too young to apply for work but he had taken
along his older brother's birth certificate which netted him a
job as a 'clippie', a bus conductor. Gerry was one of thirteen
children, of whom seven boys and four girls survived. 'My
father had a thing about the number seven,' he says. 'He had
seven fields, seven cows, seven pigs and seven sons!'

Gerry's father, Johnny King from Dromin in County Louth,

Gerry King competing in the Senior Horse Class in Ratheniska, 2014

farmed 26 acres, a big farm in the fifties, and his sons started on the tractor when they were teenagers. But little did Johnny realise the talent he was rearing, and that Gerry would become an undisputed champion, winning more All-Ireland senior titles than anyone else – a total of twelve – just pipping the late Jack Halpin who held eleven.

Now aged seventy-four, Gerry remembers starting to plough and competing as a minor in 1958. 'A local man came coaching us, at that time four of us were ploughing – Jack, Joe, Francis and myself. Once, all four of us took part in a Louth ploughing match in the late fifties.'

Gerry says there are just a handful of local families who compete at the local matches today, the McKeowns and Brendan Byrne, and the Kings. 'The others are all coming from outside and there's quite a few from Northern Ireland.'

It's a cold winter night and while Gerry's wife Mary lifts scones and brack from the Aga, Gerry tells me how they met.

'It was St Stephen's night and our first date turned out to be a Pioneer social. Dermot O'Brien was playing and we started going out. But then I had to go to England to work but I'd

Ploughing a straight furrow

come home in the winter pulling beet and then I'd stay for two months and do my courting and head back again!' The couple are now married fifty-two years and have five children.

Gerry had a five-year plan, to earn enough money to buy 15 or 16 acres of land, but it took him a while. 'The first bit of ground I got was in 1973!'

Off the farm, Gerry was erecting hay barns. 'I realised there was concreting to be done in them, so I bought an old Nuffield tractor and a loader and started mixing concrete and building silos. I started the business the week Nelson's Pillar was blown up in March 1966. I was on my way to a job in Roundwood in County Wicklow at the time.'

Strangely enough, Gerry was a latecomer to horse ploughing and only bought his first pair of horses, Jack and Barney, twenty-six years ago in 1991. Then he had a heart bypass in

David King, helped by his father and his brother Gerard, battles the elements during the Championships, 2012

the November of the following year, but was back training his horses on 11 January. 'I was always fairly fond of horses,' he says, 'but my son Gerard was into the vintage tractor ploughing and we decided to switch to horses because I really wanted to plough with horses. But Jack and Barney had problems. One suffered from ill-coordination and I sold the other fella as a pet.'

After his bypass, Gerry started working with two young horses and he won his first All-Ireland in 1998; his son had already shown the way the previous year, picking up the Under-40 title. Gerry kept winning and so did his sons, Gerard and David – the three currently hold thirty-four titles between them. Gerry says he has dedicated the achievements to his own father. He too was a ploughman in his time and won in the

1934 Louth–Meath contest. 'That gives me bragging rights. It was also the first competition held in the northeast since the troubles. There had been no local matches for almost twenty years.'

Gerry has six horses now, they are all the French breed of Comtois. There's Tipp and Richie that came from Mitchelstown, Jack and Jill from France, and two others that have never really been named! Then, there was an unexpected foal last summer.

'A bad horse is as dangerous as a bull.'

> We didn't know the mare was pregnant, the sire was only half her size! But the vet was examining another horse's teeth and he noticed the mare was pregnant. This was only a couple of months before the Ploughing, so I had to borrow another horse. People often say you must love the horses but there's good ones and there's bad ones. I like a quiet horse, a horse that kicks or is stubborn is nasty, he's dangerous, he'd kick you, he'd kill you. A bad horse is as dangerous as a bull.

Gerry King leads Jack and Jill to their plot at the Championships, 2016

Gerry has another claim to fame, he makes the most artistic and beautiful mould boards for ploughing. I first saw his work at a match when one of the competitors drew my attention to the intricately engraved images of a wheat sheaf that Gerry had made.

Although Gerry ploughs with horses, he has a fascination for old tractors and has been known to search not only the ditches and byways of Ireland for them but also the great prairies of the United States. His interest began in 1967 when he was building a lot of new sheds.

The young farmers were all coming out of colleges and the whole thing was to knock the old sheds and build new ones. And some of them had old machinery and all they wanted was to get rid of them. So I started to buy them. They are mainly Allis Chalmers tractors, but there were some real old ones and collecting those vintage tractors kind of got a hold on me. The oldest I have is a 1912 old American prairie tractor.

Then the big adventure started. When I was coming near retirement, as I thought I was – but I'm not retired, even now – I always wanted to go to America and see how things were done over there. You'd always hear people coming home and talking about America. But it wasn't cities that were in my mind. I travelled all across America looking for old tractors, old prairie tractors. My grandson Conor came with me. He was only ten when we went first.

'We'd travel across the prairies and try and find these old yokes.'

Gerry went every year with a couple of friends. 'We'd travel across the prairies and try and find these old yokes. I'd start on the internet and I'd find a local collector in the Great Falls region of Montana and I'd go out and have a good look around, they like talking to Irish people, you know, and we'd get to know them fairly well and we'd buy the old tractors. Then we'd organise a container and ship them home.'

And Gerry was not averse to hunting down the various bits of individual tractors. 'I got most of one about eighteen miles from Great Falls and the rest of it, the same tractor, way down at the bottom of Idaho, six hours' drive away. I criss-crossed Montana, on down through Wyoming, over the hills and on into Nebraska. I crossed Nebraska and up through the two Dakotas.'

Today, Gerry has about fifteen pre-1920 tractors and forty

Two of the tractors from Gerry King's collection

'middle of the road' vintage machines from the 1940s and 1950s, and they have become a tourist attraction in their own right. 'People come to see the collection,' says Gerry. 'The odd busload arrives and the nice thing is that the visitors leave a donation for our local Ardee hospice.'

Cork

County Cork is one of the most active ploughing counties in Ireland. It holds sixteen ploughing matches – eight matches in East Cork and eight in West Cork – and both sides send contestants to the Nationals. My first port of call is a group of vintage tractor men and horsemen.

'Since I was very small, I had a big interest in smoke and wheels and noise and smell.' That's Philip Cotter of the Vintage Tractor Association. He himself made the big switch from horse to tractor sixty years ago, when he was just nine years old. 'We had a horse at home, a mare called Moll, a very diligent, hardworking mare. We had no car, we had a horse and trap so she had to be kept shod. I didn't have any interest in horses but any

Philip Cotter competing in the Imokilly Ploughing Match in East Cork, 2016

time a car or a tractor would pass the road, I'd be straight out to see it.'

Philip's father Denis got the farm from the Land Commission in 1942.

Anyway, my father asked me to take the mare to the forge to be shod. He put me up on her back and I went about a quarter of a mile, and there was a buddy of mine who used to go to school with me, this was of a Saturday, and he happened to be out on the road, and I thought, this fella now is going to blaggard me, so before I got to him I got off the horse and I walked past. Then when I went to get back up on her back, every time I went to make a move, she moved, and so I walked to Donaraile, the four miles, with the horse beside me.

The mare was shod and when the blacksmith was finished he said to me, 'You'll ride her home, of course,' and I said, 'I will.' I didn't have the guts to say I was afraid, so he put me up on the horse, and as I was coming out of Donaraile, there's a housing estate, about twenty houses, Rockfort Terrace, just another quarter mile out of the village, and there was a number of fellas that were going to school with me out hurling. I was afraid of them then, and I jumped off the horse again, and I couldn't get back up. So I walked home the four miles. And that was my total association with horses after that!

Philip went on to become a tillage farmer – 'There were no cows because I didn't have a milk quota, I was out of that era' – and he told me the story of how he bought his first tractor.

I was working in the sugar company when I was in my twenties, and I went to the bingo one night in Mallow and I won £75 of a Friday night. The following Monday morning, I was in a farmer's yard and there was a Ferguson 20 tractor in the yard. I got talking to the owner and he asked me if I was interested in it. I told him I might be, and before I left the place I had bought the tractor for £50 and I had £25 left and that was in 1968. I spent that £25 on a battery. This thing was kind of parked up. He only bought it for the steerage hoeing and this was in the fall of the year and he didn't have use for it so I bought it and it was only, in modern technology, a bucket of scrap, but I thought it was a Rolls Royce!

When I got home, my parents were none too surprised. They knew I was a bit scattered. My mother said I broke everything she ever had, clocks and everything, trying to see how they were made. I was interested in seeing how things worked.

The tractor lasted for a year or two and, in the mid-1970s, Philip progressed to another Ferguson in better condition. Then he got a plough from a neighbour.

As time went on, Philip says he didn't have the money for the modern stuff, so he decided to stick to the old, vintage machinery and to keep it going. 'I'd take anything apart and investigate it, I broke some things and fixed others.'

Then he discovered the Vintage Ploughing Association, which was formed in 1991 in the Mason's Apron in Urlingford, County Kilkenny. It progressed to three categories – trailer ploughing, single-furrow ploughing (which was little garden tractors) and vintage trailer-mounted or hydraulic. They have about seventy members now, but they wouldn't all be ploughing.

'You'd have a lot of people who have ploughed in the past, they'd be members and go to the meetings, and you'd have people who might be interested in the future; they might have a tractor and no plough and they'd join up and they'd be looking in the direction of going in the future or a son of theirs taking it up.'

To qualify as vintage, tractors and ploughs have to be pre-1959. A lot of the machinery of this era has been pulled out of hedges and ditches on farms and refurbished. Philip says that most of the machinery made in the forties and fifties seems to hold an awful lot better than the modern stuff, so if you pull them out of the bushes, they can be rebuilt. But it takes dedication and a lot of time. 'You need a lot of patience, you might have to strip down the same item four times before you get it right. I spent a year and a half doing up the tractor I ploughed with, getting it onto the road, most of it because there wasn't enough money – I had a mortgage and a wife and family, any few pound I'd have, I'd spare it.'

A lot of the machinery of this era has been pulled out of hedges and ditches on farms and refurbished.

What makes the tractors particularly expensive is the fact that most of the parts have to come in from England, not a lot is manufactured in Ireland, so between exchange rates with sterling and transport costs, it's expensive enough. The most popular tractors are the Ford and Ferguson and John Deere, they were the older ones, and the Ferguson 20, the 35, the Dexters, the Internationals, the David Browns, they were all of the era. Philip has the oldest working tractor, a Ferguson – was found up the midlands about ten years ago, it dates from around 1946.

Philip has ploughed competitively, but he didn't bring home any silverware. 'The nearest I got to it was fifth in the Two Furrow Mounted in the All-Ireland Vintage in 2014.'

Sixty-five-year-old Willie Stokes is another Vintage ploughman who farmed in the village of Liscarroll. His next-

Willie Stokes

door neighbour has now taken on his farm because 'he's big into milking cows and needed extra land when the nitrates came in. He asked me would I let it to him, otherwise he'd have to reduce numbers, he has it now full time.'

Willie grew up on a family farm and spent his own time farming tillage and contracting in partnership with his son Mark and his brother John.

> I was on a tractor from knee-high. My grandmother used to give out! I actually have a photo of the first tractor I drove. I'm sitting on it, and steering it – I was only four. My grandmother saw from the yard and there was murder. I shouldn't have been let onto it. I've a lump here on the back of my head from when I fell off a little grey Ferguson, I was about five or six then. It was a very different era.

By the time he was fifteen, he was out driving the tractor on other farms in the area. 'I wouldn't have been driving on the road, the first year I was out cutting silage, which was 1966. I was going to school at the time and, in the summer holidays, I'd be cutting silage. I left school in 1968 because my father had a car accident – 5 May 1968, I'll always remember it. It happened about a mile from the home place, he was going to a funeral. It was a bad accident and he was unconscious for seven weeks; he did come out of it, but was never the same again. I came home from school and took over the contracting, I was running it at that stage.'

He tells me his mother was also in the car accident and she was unconscious for five weeks afterwards.

Willie says there was 'fierce' pride in the farm work in his father's time.

You had to get it right, particularly when you were alongside a road. Opening drills would be more particular. They'd have to be straight. I remember, we had three or four men working at home, and none would open drills with my father, unless they had a harrow in the field as well. When they opened the first one, he'd harrow it out again if it wasn't straight. This was for turnips or carrots, beet even. It didn't make a difference to the crop, it was only appearance and pride. At that time, you had an awful lot of people cycling and walking to mass and that, and everybody knew one another and they all discussed what was happening, and all that. If there was some good person, they'd automatically compare to him. And you'd know it within hours of it. You'd just get on with it. It was just slagging, 'the sun was in his eyes', or something like that!

'At that time, you had an awful lot of people cycling and walking to mass and that, and everybody knew one another.'

Although he never ploughed competitively, Willie had a great interest in going to the National Championships, so much so that when he was in boarding school in Waterford in the early sixties, he'd go for the day. 'They picked me up and dropped me back that night. I'd be picked up in the morning at nine o'clock and brought back again at six or seven in the evening.'

'Footing is illegal. Footing and handling and fixing – positioning a sod to make sure it doesn't blow away with the wind.'

He has been involved in his local club in Kilbrien, which was formed in 1982. He is the time-keeper and rule enforcer and has many a story about the arguments that break out over the rules.

'Footing. Footing is illegal. Footing and handling and fixing – positioning a sod to make sure it doesn't blow away with the wind. But if it's a big field and forty competitors, while the judge is looking at one, thirty-nine are free! It's one man's opinion against another.'

Philip was reminded of Anna May McHugh's famous quotation. '"Any man who comes second in a competition never agrees with the judges,"' he says. 'If he's beaten by one or two marks, he feels a bit done.'

The judges have the final say, and there's never a temptation to help your own county men win.

You'd always have someone out supervising and watching what they'd be doing and shouldn't be doing. But I'm supervising since 1991 at national level, and you'd always know when a competitor is doing something dodgy because his helper, standing on the headland, would start watching you, to see were you watching the chap down below who is up to something.

And mobile phones are not allowed. There can be no communication from the headland. I remember one time a chap ploughing at a local match down near Youghal, and he's from another county, and he came in and he was working away and he ran down and was doing some repair work by hand and I said, 'You know, Paddy, you're not allowed do that?' And he said, 'Oh you can't *walk* in the ploughing – but you can run!'

And there's another individual, up the country, not in the Vintage Class but in the Conventional Class and he used to use cameras on the back of the boards and have a little screen inside in the cab of the tractor. It was legal at the time, there was no one stopping it. It was eventually stopped. An awful lot of people that are ploughing are not farmers at all, they're using it as a diversion and a hobby. They're mechanical engineers and they have workshops and they're trying to beat the system with mechanical devices and they're constantly thinking of adding on bits and pieces. In the Vintage, we're not allowed have anything, no add-ons at all.

Another Vintage ploughman, Connie Hartnett, is seventy-one now and farms near Banteer in North Cork. When he was growing up, they used horses for all the farm work. His earliest memories are of hand-milking cows when he was ten or eleven. He had two brothers and four sisters but they had no interest in farming. So Connie left school at fifteen and went to work on the farm. His father was very interested in working horses. 'He used to train them as two-year-olds and get them working, he'd be selling the trained working horses, there was a big market for it at that time. One spring, he sold two working horses, and bought two horses that had never been trained, and he trained the two of them together and that was in 1967, and in 1968 they competed at the National Ploughing Championships at Rathcoole near Banteer, and he got a prize for the best pair of working horses.'

'He wouldn't give her to anybody because she was so diligent and loyal all through the years.'

Connie's father was very attached to his horses and when one of them died on the farm, 'Somebody came to buy him for the knacker's yard, and he wouldn't sell her. We had to dig a big drain to bury her with shovels and pickaxes. It was a day's work! He wouldn't give her to anybody because she was so diligent and loyal all through the years.'

All the men agree that it's very important to talk to your horses. 'To train horses to work together it is very important that you'd be able to talk to them and tell them to go steady or slow down, you'd have your own language – "Steady now, steady now, up up."'

There were ploughing competitions in Banteer in the thirties and forties, but there was a lapse then from the late forties until it was revived in 1964. Connie remembers a meeting in the hall in Banteer where they were talking about running a ploughing match and saying they needed a book for their fiftieth anniversary in 2014. He pulls out the beautifully produced anniversary book with photos on the cover of the four men who had been involved in the first match fifty years earlier. 'The four of them were involved in the running of every one of the matches for fifty years,' he says. 'That's myself on the bottom left, then Donal Lehane, Dermot O'Flynn and John O'Connor and that's a monument that we erected in Banteer.'

Connie Hartnett

Connie competed in the ploughing in the early days at the local matches 'but I hadn't time for them so I said I'd give up and stay on the organising end of it. And I've done that since. That was the late sixties, I haven't competed since then. I've never competed in the Nationals.'

Connie became Chairman of the Banteer Committee, Vice-Chairman of the County and EPA Director for the county. He talks about the responsibility of getting the team ready for the Nationals each year.

*Banteer Ploughing Association's
Fifty Years of Ploughing*

We had the county ploughing match in 1970 and we were responsible for the team going to the Nationals in 1971, it was held in Finglas. We booked a hotel, the Sunnybank Hotel, booked in all the ploughmen and helpers, and we had to be there to meet them when they arrived. Four of us got on the train in Banteer, on a Tuesday morning, went up, booked ourselves in to the hotel, got a taxi out to the site and came back in that evening to meet the competitors as they were arriving but there was no sign of the horsemen.

At the breakfast the next morning, Jerry Horgan and Gerry Delaney, both competing, came in for their breakfast. Now at that time, the ploughing used to be the last week of October. It was a dry, cold, frosty morning, and Gerry Delaney came in in his short sleeves and I said, 'You're warm yourself this morning', and he said, 'We slept in the lorry last night and the horse pissed down my coat – I couldn't wear it!'

John Horgan, Jerry Buckley and Jerry Horgan at the Nationals in Finglas, 1971

Johnny Horgan won the All-Ireland anyway and we had our ticket to come home on the train. But we decided it would be more fun to go home in the lorry with the boys! There were two pairs of horses in the back of the lorry, the two Horgans, Thady Kelleher, Gerry Delaney, Jack O'Connor, Jeremiah Leary and myself. But coming down through O'Connell Street in Dublin that evening, Johnny Horgan was drinking a pint bottle, and the minute the bottle was empty, he thought he was at home and threw it out the back of the lorry! And a line of traffic coming behind us on O'Connell Street! You could do things that you wouldn't do now anyway!

As an aside, Connie says the Horgan boys used to make their own poitín. 'Gerry and John from Ballinagree. People are still making it down there! There are pockets down there where they'd drink it breakfast, dinner and supper. They don't bother going to the pub.'

One of the local highlights was when the National Championships came to Banteer in 1968. 'They were looking to break a world record, 3,000 tractors working together,' Philip says, 'and they finished up with 4,752 tractors, and it hasn't been broken since. It was a massive field and they had to be all ploughing with some kind of an implement behind them. It's a Guinness world record.'

'We do ask ourselves about the future,' says Connie. 'How much bigger can the championships get?'

'At the moment it's the biggest event in Europe,' Philip says. 'And I've been at a few world events in France and England, and, by comparison, they're only a speck. One of the reasons Ireland has been so successful is that it's kind of at the end of the season and it's nearly a national holiday. For people working on the land, they probably wouldn't bother taking any other holidays and they'd make that their national holiday.'

'You'll notice even in bad weather, when the place is full of muck, people are still smiling,' says Willie.

I ask them about the romances cemented at the ploughing – or whether that's just a rural myth.

'We do know of girls who were ploughing and they might have a helper with them and the helper might have finished up marrying the girl three or four years after,' says Philip. 'It might have happened because of Macra na Feirme, he happened to be giving her a hand in the fields and they eventually married!'

They do name names for me – but I'm not allowed to write them down! So I ask if they ever met anyone themselves.

'Well, I was married the year I got involved in the ploughing!' says Willie.

'I met several people at the ploughing,' says Philip. 'But sure you'd have to meet them because if you were put supervising on the Farmerette plots, you'd meet quite a few women! But I got snared before that! You'd meet a lot of farmers' wives, they'd be staying in the hotel and that, so you'd meet wives and daughters. They used to have a great dance for the Queen of the Plough but it's after fizzling out. There were no inhibitions then, no border lines, they were great.'

Thady Kelleher Country

In the spring of 2017, I travelled to East Cork, a place where the rivers Allow and Dallow meet en route to the Blackwater. It's the place where, legend has it, the last wild boar was slain in Ireland and Ceann Tuirc – or Kanturk – is translated as 'boar's head'.

But I'm here because this is Thady Kelleher country, and just outside the town, on the way to Banteer, there's a larger than life statue of Thady and his plough. This is the man who won over forty county titles and two world championships, a giant of the furrows. Just a month before his sudden death in 2004, he was among the prize winners at the Nationals in Tullow, County Carlow, where he was placed second in the Senior Horse Class. The man who came first that day, his archrival and friend Gerry King from Dunleer, County Louth, said, 'Anything I learned about ploughing I learned from Thady.' And Anna May McHugh added that Thady was 'a true mentor in ploughing circles'.

Anything I learned about ploughing I learned from Thady.

I was on my way to the Kelleher farm to meet Thady's brother Dennis. I got lost several times but eventually arrived at the right boreen, the right farm, the right dogs ... only to be engulfed in what I can only call an impromptu hooley.

Thady Kelleher's statue, Kanturk

THADY KELLEHER
1935 - 2004
WINNER OF WORLD
AND ALL IRELAND
PLOUGHING CHAMPIONSHIPS
THADY KELLEHER
From the fields of Duhallow
You bring back the waves of our
yesterday
You feed the dreams of our tomorrow
Peaceful and strong is the
ploughman's way
JOHN DILLON

Instead of only meeting the illustrious Dennis and his lovely wife Judy, he had assembled a who's who of everyone involved in ploughing in the region, and there were about twenty-five people in their kitchen, which was also filled with platters of sandwiches, home baking, delicious-looking tarts and teapots that never stopped pouring. They were all there because they had one thing in common: an obsession with ploughing.

Dennis is one of the eleven children of Dennis C. Kelleher and Maria Kelleher. Only four – Eileen, Margaret, Anne and Dennis – survive. Six of his sisters – Anne, Peg, Mary, Sheila,

Eileen and Kay – all trained as nurses. His late sister Sheila wrote a book of memories after Thady's death and she recalled the long-ago days on the family farm and the threshing, when their uncles would come from Tooreenbawn in Millstreet.

> The thresher rolled into the yard early in the morning or the night before. The mother's friends would arrive from Gurranduff about eight o'clock in the morning. They made a panger for the *daornéal* – the panger was a thick, very long *súgán*. There was straw put on the ground first and the panger was placed circular on the straw for the storage of the grain during the winter. The grain was levelled as it was put in. When the work was complete it was cone shaped and was water and rat proof. Of course there was no school that day – that was a foregone conclusion.

Cork's 1954 All-Ireland-winning team – Ted Keohane, Jerry Horgan and Donal O'Donovan

Another of Thady's sisters, Eileen, describes the operation as 'like twisting a big wet bath towel, wringing it out and the *daornéal* itself was a large cone – it could be up to 8 feet in size – used to store the grain in the field, it would be left in the field'.

In her book *Star of the Plough*, Sheila also tells the story of fundraising in a small parish when Fr Tangney, the parish priest, put a levy of £5 a cow on each farmer of the parish to pay for renovations to St John's Church in Dromagh. There was a parade as well and she tells how a few witty boys dressed as mourners and carried a coffin on their shoulders with a placard which read, 'Death due to shock of £5 per cow!'

As we sat around Dennis's big kitchen table, Dermot Flynn said he could remember the first ploughing match that Thady ever went to in 1958. 'It was back in Timoleague. We used to be ploughing locally all right but we decided we would go to Timoleague. We said we'd get a pair of horses there and we did. We had an old Prefect ZT4905, and there was no hitch on the car, so we put one on the night before and we took off the next day. Got a pair of horses and everything was grand, then about halfway through, the horses failed to go any further! End of story!'

That was the start of Thady's career in 1958 but from there onwards it took off. The famous ploughmen of the time were Jerry Horgan and Murty Fitzgerald, Ted Keohane and Mossy Sheehy and 'you had to fight for your place'. Then Dennis came on the scene.

There has always been great rivalry between Cork and Kerry, but out of the furrows they're all great friends. Sonny Egan (the blow-in from over the Kerry border) says it wasn't about the competition. 'When you were out there you were trying to do as good a job as you could in the day and then it was up to the judge to decide but everyone was concentrating on their own job as good as they could. Things wouldn't always go right, you could have a break or stones in the plough. And there'd be funny things too, you could have helpers that might go to the pub and never come back!'

There was a lot of chat about 'the woman of the house'. Sonny says she was looked on 'as a brilliant woman if she had the tackle and the harness and the brasses all shined up. And she'd bring the food for the ploughing, women were very important. The men wouldn't be there without them, and the men were more afraid to come home to their women if they didn't do well in the ploughing than if they lost an All-Ireland.'

Dennis Kelleher directing Sonny Egan at the plough at the Nationals in Athy, 2009

The celebrations that greeted the victorious ploughmen went on for days. When people arrived home there'd be flaming torches and they'd be carried into the village. It would be spoken about days before and days afterwards and the men at the time, the Horgan brothers, Dan J. Mahony and Moss Trant and John Joe Egan, became heroes.

But on that February day in the Kellehers' kitchen, among the people who ploughed alongside Thady and who knew and loved him best, the memories were coming out of the woodwork. And no surprise, we were surrounded by his trophies, including the beautiful silver harness he won when he came first in the Worlds in Horncastle in England in 1984.

His sister-in-law Judy remembers the time. 'There were three of them travelled, Dennis and his friends Tommy Reilly and the late John Joe O'Connor. It was their first time out of Ireland and they went by boat. They didn't have horses but they decided they'd go anyway and borrow two horses. As they were passing through Portlaoise, the exhaust went on the car and the guards came out to investigate when they heard the noise.'

His sister Anne remembers it too: 'They left about three o'clock in the morning. On the way, they called in to Eileen [another sister] early in the morning. I don't know what day of the week it was but they realised they were travelling to the UK and they had no sterling! Eileen knew the local bank manager and, despite the fact that the bank was closed, she rang him at home and he agreed to meet her early in the morning outside Mitchelstown to hand over the cash! When they got to the North Wall in Dublin, they met up with John Colleran and the other ploughmen from around the country who had qualified for the championships and they travelled together on the ferry.'

Also competing in Horncastle was Gerry King. This was the first time the two got to know one another. In his tribute to Thady, Gerry remembers.

We were in a public house the night before the competition and I happened to overhear an Englishman saying, 'Did you see the Irishman's plough? Looks like he's just pulled it out from the hedge – he shouldn't be allowed plough with that thing.' Thady didn't hear the remark but turned out the next day to plough his plot wearing the same cap and double-breasted suit that he was wearing in the pub the night before. If his plough didn't look like much, Thady was certainly the best-dressed ploughman on the field that day. And he was the outright winner!

'And when he won,' says Judy, 'we had a bonfire below. Everyone came, the whole place was full. The singing and dancing went on all night long. Then Thady came in and danced over the brushes, that was a fierce night.'

Sonny Egan remembers the celebrations in Abbeydorney when their champions came home. 'They were heroes,' he says. 'It was great to think you'd rub shoulders with the likes of these men who won All-Irelands at the Ploughing, for me it was like going to the All-Ireland football final and playing with Mick O'Connell around the middle of the field.

'Tis a religion, 'tis a tradition and most of all, I suppose 'tis the enjoyment. To the ordinary man, sitting on a plough looks awful easy, no bother to him because it's going right but the man that's behind it has to adjust it for every different scribe, for an open he also has to have the horses trained that they won't be walking up on top of the scribes.'

But even at local or national level, there were huge preparations. Anne and Eileen remember the fuss and the mad scramble the night before a ploughing match. One time, they couldn't find the feathers for marking the plot. So they had to try and find someone who kept geese, it wasn't just any old feather, it had to come off the wing, there was more of a flip in it.

Then there was a big box of tools to be checked and then the horses had to be shod and maybe the collar would need to be repaired by John, the harness man.

It was an all-nighter. Thady made holes in the side of the travelling box to rope up the plough and they had to take some drink. I was reliably informed that this was holy water!

Anne says one year they went off to an All-Ireland and forgot part of the plough. But they managed to contact someone who hadn't left yet and got it brought out to them.

> *''Tis a religion, 'tis a tradition and most of all, I suppose 'tis the enjoyment.'*

Jerry Horgan (left) stops for tea at the Nationals, 1957

And there was a story to every ploughing match. Thady and Dennis had a horse box in 1994 and 'they wouldn't be interested in going in anywhere to sleep, they'd sleep anywhere rather than going in to a hotel, so they decided they'd sleep with the horses in the box. There was a barrier in between the horses and themselves at the back but they never knew there was a fall in the floor of the lorry and the horses peed and drowned the men below. In the morning, they got up and they were all wet and they put their clothes out on the bushes to dry them. They were all yellow the shirts and they had to wear them.

These men had a following and the young would-be ploughmen watched them closely. They would be fascinated to see how they controlled the horses and the plough. But by the time the younger men were old enough, the horses were dying out and tractors were taking over. But the decline was halted in its tracks because, despite the efficiencies of the tractor, the horses are part of their tradition, as important as football or music in the community. And they also realised that good ploughmen were respected in the community and had a special standing.

There are five horsemen gathered round the kitchen table: Tim Lawlor from Rosscarbery, J.J. Delaney from Ballinagree, John Sheehan from Barraduff, John Donovan from Derrybawn and Sonny Egan who had crossed the county line from Abbeydorney in County Kerry.

J.J. Egan competing in the Nationals in Screggan, 2016

Tim won the All-Ireland in 2014 with his horses Freddy and Larry. He told me that getting ready is hard work, troublesome work, you have to get a bit of practice in first if the weather is good. If the weather is bad, you can do nothing. Tim has been working with his partner Johnny for the past fifty years. It all began when Johnny used to take Tim to school in the mornings, he worked in Clonakilty and gave him a lift every day.

J.J. started ploughing when he was thirteen but didn't get a tractor until he was eighteen, and was delighted to get rid of the horses. 'I ploughed in a few local matches. When Johnny Horgan died, I was asked to take over and go back to the horses again and I decided to give it a go again. In 1995, I won the senior at national level and my young fellow won then in the Under-40s. In 2016, he ploughed at the Worlds in York and came fourth. So we're happy with the horses.'

But J.J. also spoke of something you don't hear much about, the sheer loneliness of the ploughman. 'Ploughing on your own can be a boring old day. For an acre ploughed, you'd have eleven miles walked with the horses! But there are compensations – you see everything, it's the nearest thing there is to God, the crows are following you, picking, and the seagulls flying very close. It reminds me of the poem Gray's 'Elegy' where the ploughman "plods his weary way".'

'You'd talk away to the horses all the time, good, bad and indifferent. When the man catches the ropes behind the handles, the horses would know if there was a different man there.'

The ploughman has a great affinity with his horses and his plough. J.J. says his horses know the sound of his voice, and understand what is said to them. Dennis Kelleher joins in. 'You'd talk away to the horses all the time, good, bad and indifferent. When the man catches the ropes behind the handles, the horses would know if there was a different man there.'

'Horses are great company,' says John. 'You'd enjoy being out with them. You see foxes and badgers and mink and even deer.' John has been ploughing since 1955, but he's discreet about his actual age. 'That's a guard's question,' he laughs.

When I ask if the ploughing will survive, Mary McSweeney is quick to say yes and talks about how much she looks forward to it each year.

'Why wouldn't it survive?' asks Dennis, indignant that I'd ask such a question. 'It might change a bit but it will survive. The ground must be ploughed every year anyway. And there have been changes. When we started ploughing first at the All-Ireland, 'twas all lay ground, it has changed a bit, some of it is stubble, it could be because the big machinery companies out in Europe have designed these ploughs to plough the stubble, and forced them onto the Irish people, that may have forced the change.'

Dennis remembers his own first competition in Banteer in 1967 when he borrowed two horses from John Joe McSweeney. They were two good horses, Arthur and Molly. 'I should have had Jerry Horgan beaten, but he was too cute! It was a very frosty day, four of us were there, it was so frosty you couldn't plough in the morning without skidding. But Jerry was cute, he waited and he got Murphy's horses in the evening when the ground had thawed a bit. He ploughed so many sods and it got pitch dark and they were all gone home. And he won! Sure he'd win anyway in fairness. I came second then.'

It was time for apple cake and more tea and we moved on to the tractor men and Jeremiah O'Sullivan who told me that he took up ploughing because 'a few of my neighbours were good at it, so I tried it'. Competitively, he claims he had a bit of hard luck. 'I met a rock in the furrow. I was winning it well up to that. But I still love ploughing. It's nice, I just like it, that's it. I like being out with the crowd, the show. In the All-Ireland, there's plenty to see, I'd be watching the ploughing a lot.'

Dermot Glynn says he fell into ploughing by accident. He was serving his time as a mechanic in a garage at Banteer that sold farm machinery. When someone bought a plough or a tractor, a staff member was sent out to get them going and Dermot was duly despatched to the local customers. He says that in the beginning he knew less about the machinery than the customers did but 'eventually I got myself going'. Dermot took an interest in the ploughs he was selling, and started to tackle up a plough and get a loan of a tractor for the day and in

1987 he went on to win an All-Ireland.

He remembered Thady's criticism when he'd look at a few furrows. 'He'd ask, "Where were your eyes when you ploughed that round?" He was some man with the ploughing, it was determination and practice. You always need someone to help you, to come out and tell you where you're going wrong. I remember well turning boards out in that yard. Many a night, we were here lighting fires in the middle of the yard, turning the board and going back the next day to try it. You'd keep at it till you got it right.'

Jerry Horgan at the Nationals in Finglas, 1971

Thady was a big, strong man, he said, 'but a rogue, a bit like Dinny!'

'Isn't that an awful thing to say in my house?' countered Dennis.

There was Thady's driver, Michael Herlihy, who used to get up very early to load the horses and drive them to competitions – at one time he led the horses too.

And John Sheehan who won two All-Irelands met Thady in Muckross Traditional Farms. 'Thady was there for a short time training me. That wasn't easy! He was a legend, I was fierce lucky to have him training me.'

I asked the group about the social side of the ploughing, the dances, the fun and the possibility of romance. But they were all adamant that, while others were cavorting and having a good time, they were as well-behaved as students in an Irish college. I pointed out that I had attended Ballingeary Coláiste na Mumhan and begged to differ, remembering the midnight feasts, the Céilí Mór, the Bean an Tí and the 'fine things'. But John said the only woman he ever met at the ploughing was Nell McCafferty.

Work done, interviews over, it was time for recitations and dancing, and Sonny Egan entertained us with a gravity-defying, trouser-threatening version of 'Rock Around the Clock'.

I left Banteer and the Kelleher farm after many hours, grateful that I had stumbled into the gritty, caring, warm embrace of rural Ireland, that they had allowed me to be a part of their family and friends and neighbours in the parallel world of the ploughman.

Thady Kelleher

'I was fierce lucky to have him training me.'

Zwena McCullough and the 'Black Sod'

Zwena McCullough sitting on her All-Ireland-winning plot, 1996

'Zwena, you can't plough!'

'Why is that, Thady?'

'You can't drink, and if you can't drink, you can't plough!'

That's what Thady told his most unusual ploughing ally, Zwena McCullough, the first and only woman to plough against men in the Nationals. She and Thady both took National titles in 1996 and ended up ploughing in the World Championships as a result.

Zwena has a fascinating background, having grown up at the Hydra Farm. As she tells me, 'It was the first place in Europe where there was treatment for cholera and TB. It was built in 1856 by Dr Richard Barter. My father, Hugh Quigley from Glasgow, was also a doctor and he came over to Cork with his sister on holiday; he was more or less retired then and they stayed at the Hydra. While he was there, he was invited into the office and they offered it to him for sale and he bought it!'

Zwena spent a lot of time abroad, but when her father died in 1983 she came home and stayed.

We're in the tea rooms in Cork's Imperial Hotel and the first thing Zwena tells me is that she doesn't like the term *farmerette*. 'It's like calling someone a *womanette*, a little bit of something. They should change the title! These girls are very talented and they work very hard and I think they have earned their stripes and I think to call them farmerettes means you haven't quite made it!'

'It's like calling someone a *womanette*, a little bit of something.'

I'm here to discover Zwena's story, and how she ended up competing against the men, but first she tells me about her initiation into the world of competitive ploughing. It was the 1990s and she had been ploughing with tractors, driving for contractors and cutting silage.

> I love machines and the more power the better. Then, in 1995, I went to a ploughing match with the silage crew and I could not believe it when I saw the horses ploughing. I saw these two Clydesdale horses coming up over the hill and ploughing and I fell madly in love, I was just completely besotted by this! I couldn't leave the site, I went down and asked several of the people if I could plough. And, of course, they laughed at me and said, 'Of course, you can't, so get lost.' But I wasn't put off in the least, all I needed was one person to say yes.

Then Zwena met Bertie Hanna, 'this absolute character'. 'He said I could in the complete craziness of the moment! Bertie lived in County Down but J.J. Delaney said he'd take me on and he taught me how to plough.'

Zwena has great memories of the local events where she first competed.

> You would arrive at the competition and you'd book in first, then you'd have a glass of whiskey with all the others before breakfast and then you'd plough for four hours. Then, that evening we'd head off for supper to a hall somewhere and we'd have this lovely bacon and cabbage and lots of good food. I remember a New Year's Day ploughing match in Skibbereen. Alan Barry has a pub there and he's got all kinds of old-fashioned stuff and he does all kinds of festivals and he had organised a festival on New Year's Day and we were invited to go down. Coming home, I had to travel 70 miles with the trailer and it was very, very icy and snowy and when I was driving along the little road, the horse box started sliding into the ditch. I had the two horses in it, and I just kept going hoping it would get itself right and it actually did. It's not an easy life, it's not a life for many people. But it's so exciting!

'It's not an easy life, it's not a life for many people. But it's so exciting!'

1996 was Zwena's year, when she won the Under-40 Horse Plough in the Nationals at Oak Park in Carlow. 'I had only been ploughing for a year, but I entered the Nationals anyway. I had taken part in every competition that winter, learning with J.J. And the very same plough won in 2016 in the Nationals with J.J. Delaney's son Jeremiah. I won the Under-40 title and Thady won the Senior one and, because it was the World Championships that year, we went on to plough for Ireland on the fourth day in the Worlds. Thady came first and I came second – six nations ploughed, and Ireland came first and second.'

The first time Zwena competed in a horse-ploughing match, Ballacolla, 1995

Thady's sister Sheila Foley recalled that Thady was delighted to win the award. He said himself, 'I was hopeful going up as I knew the level of competition was very high. I was relieved to take the All-Ireland title and I just did my best in the international event. It was a great feeling to win.'

Zwena is the only woman to have competed directly with men. What was their reaction, I wondered. 'Fantastic! Everybody was absolutely wonderful. The late Connie O'Brien, who knew ploughing inside out, said to me when I walked into the tent, "Ah, there's only one way you can go now and that's down, so don't get too excited!" I said, "Connie, today I'm not listening to you!"'

After their great win, the pair rode on buses through towns and villages and 'celebrated endlessly, it went on for days, weeks'. Zwena continued to compete at local level but she never got to the Nationals again. 'That was my one glory day!'

Zwena says she ploughed in the very best years when all the most wonderful people were there 'and a lot of them have sadly gone now'. But there's one thing that still bugs her – the end of the 'black sod'.

I am still devastated about it. At the end of your ploughing when you had finished, there was a thing called the 'black sod'. You came in with the plough and you ploughed down very deeply and you pulled up a sod that wasn't there, it was an imaginary sod. When you looked at it, it was just a flat thing, but you

J.J. Delaney and Thady Kelleher celebrating Zwena's win, 1996

pulled that up and it was the hardest sod to pull and you got fifty points for it. I loved that sod more than anything else! I would do well on it but they took it out of the competition altogether because they felt it was too hard for so many people.

After the championships, Zwena penned a poem in honour of Thady. Here's an extract:

This legend of a man, stood strong above the land,
His eyes fixed firmly on the ground,
With determination in his heart and skill in his hand,
He grasped the handles of the plough.
Slowly the horses turned with their breath like fading clouds,
In the cold morn, the plough slipped through the ground,
Turning each side and delighting the crowds,
The black sod was as straight as a rule.

But Zwena wasn't the only novice whom Thady helped to train, he was well known as a mentor for many others and indeed, before he died, Thady was coaching a young man from East Cork, John Sheehan. After Thady's death in 2004, his brother Dennis coached John for the Senior Horse National Championships and he won the title in both 2005 and 2006 using Thady's plough and Thady's horses, an achievement of which the Kellehers are extremely proud.

Michael Walsh

'The ploughing is fierce important to West Cork people, it's unreal!' The words there of Michael Walsh, who is Secretary of both the Cork West and the Bandon Ploughing Associations. Ploughing is in his blood. His father John won two national tractor titles in 1956 and 1962. But, today, Michael says, half the ploughmen in West Cork are not farmers but teachers, lorry drivers and factory workers.

Michael is passionate about ploughing and waxes lyrical about their results in 2016. 'It was a great year, we won seven All-Irelands out of twenty-five classes. Then, two of our lads represented Ireland in the World Championships in Scotland and they were placed second, Joe Coakley is European champion and Liam O'Driscoll was runner-up to Joe. We expect to sweep the boards in 2017.'

There were also plenty of Queens of the Plough from Cork, Lilian Stanley from Kilbrittain in 1974 and 1980; her daughter Marion Stanley in 1985, 1986, 1987 and 1992; and Elizabeth Lynch in 1988 and again in 1990.

Michael says ploughing is the most enjoyable pursuit.

It's great fun, there's a great social scene. You start off early in the morning, arrive in the field and most clubs would have a little caravan and they'd have tea and sandwiches, plenty of whiskey but not so much poitín nowadays. There's great banter among the ploughmen, you might throw a sod at the man working next to you. There's no doubt the ploughing will survive but I would like to see a watch on the commercial side, which is taking over a bit.

'The ploughing is fierce important to West Cork people, it's unreal!'

'There's great banter among the ploughmen, you might throw a sod at the man working next to you.'

Northern Ireland

The Northern Ireland International Ploughing Association has been in existence for seventy-four years. Compared to the NPA, it's much smaller and poorer and the members spend a lot of time during their summers fundraising to participate in events outside Northern Ireland, especially for travel to the World Championships.

Two of the most colourful characters in ploughing come from Northern Ireland. David and Samuel Gill are fifty-year-old identical twins. They were probably the youngest competitive ploughmen ever, participating in their first competition at a local match when they were just nine years old. They actually won their match in the Under-25 category. David told me, 'We drove the tractor and our father controlled the plough, he helped set the plough for us. We were only basically driving the tractor, pretty much under the guidance of our father.'

David Gill

But being identical twins has certain benefits and when they started ploughing, says David, 'I was better at making the open cut and Sam was better at the finish, so we used to swap and nobody knew anything about it! We were about twelve at this stage. We wore exactly the same boiler suits and hats, our mother always sent us out dressed the same.'

They didn't own up to the swap until ten years after the event.

I'm curious to know what it's like to win a ploughing match at nine years of age.

'At that age you don't really think,' says David, 'you just concentrate on what you're doing. There were very few spectators then, only the ploughing people and supporters; it's still like that today but ploughing people all encourage each other. Although it was competitive and you wanted to win, they would help you out and encourage you.'

The twins were sixteen when their father Raymond died. 'We thought we would carry on his dream so we started going to matches all over the country. There was a problem, though, because we were too young to drive a car and we only had a licence for driving a tractor on the road. We were going to compete in a match in Tullamore, so we drove two tractors at a speed of 18 mph from County Down all the way down. It was ten hours driving and it wasn't like today's tractors with soundproof cabs and heaters, this was only a basic cab!

'We had to stop and ask for directions near Ardee. "Do you know the road to Tullamore?" we said, and they didn't believe that's where we were heading on the two tractors.'

There was no motorway then and they had to travel on minor roads. Their uncle Bobby Moffett, and his friend Donald Gibson 'followed us in the car in case anything happened, it was probably worse for them, we were going as fast as we could and they were going really slowly and trying to direct the traffic around us'.

Eventually, they arrived in Tullamore and competed but 'we didn't do that terribly well, we came seventh and eighth in the

Under-21s'. But it was a massive learning curve because most of the competitors were using Pierce ploughs and they were using Kverneland. 'We had never seen Pierce ploughs before in our lives, most of the opposition were using them and doing very well with them.'

David and Sam's grandfather, Joe Gill, was a farmer in Crossgar, County Down, and ploughed with horses and competed in local matches. His three sons – Raymond, Winston and Martin – all took up ploughing. In turn, Raymond's three sons – David, Sam and Richie – also turned to ploughing. All three have been successful over the years, qualifying to go to the World Championships since 2000 every year except for two. But in 2005 David and Sam were shocked when their brother Richie beat them to represent Northern Ireland in the Czech Republic and finished third overall. 'Of all the things to happen,' says David. 'He's not even from an agricultural background, he's a bank manager! He was like a fish out of water! He turned up in a pair of overalls, normally he wears a suit, and he doesn't like getting his hands dirty but he went and did very well.'

David describes his own 50-acre farm in Hillsborough as a 'hobby farm' with beef cattle and grass. He and Sam are both mechanical engineers so ploughing is their hobby. David got his big break at the Northern Ireland finals in 1995, when he became the overall champion and qualified for his first World Championships in Kenya. Interestingly, it was on exactly the same site as the 2017 event. David went on to represent Northern Ireland eight times at the world event and took home the golden plough as the overall winner in 2007 in Lithuania. But he says there was a

'I worked all my life to become world champion and the only thing I can remember when I won was holding the golden plough and thinking, What the hell am I going to do now?'

sense of anticlimax. 'I worked all my life to become world champion and the only thing I can remember when I won was holding the golden plough and thinking, *What the hell am I going to do now?*'

I wondered if the Northern Ireland competitors have an identity problem. Are they seen as British or Irish when they compete abroad and win?

David is a diplomat on this one. 'Well, we get more recognition through Ireland than the UK but I see myself representing Northern Ireland. The guys I'm watching are the Republic, they're the guys we have to beat, and I hope they feel the same about us. Ireland is leading the way in ploughing and the rest of the world looks at the way ploughing is done in Ireland. If you go to the world match, the rest of the competitors are looking at the Irish because they're setting the pace. We punch well above our weight.'

Sam Gill competing at the World Championships in France, 2014

In the early years of competitive ploughing, they didn't come south because of the difficulties they encountered at the border. But now when I talk to anyone from the border counties, they stress the amazing camaraderie there is between people from either side of the border. David says he ploughs a lot in the Republic.

> 'Our young people are our future and it's important to encourage our young people in agriculture, it's the biggest industry Ireland has.'

They are so laid back and very welcoming, they're very happy to see us. At the last match I ploughed at in the South, in April 2017, it was a match in Longford and there were nineteen people from Northern Ireland competing. At a local match in the North, we would have a maximum of thirty competitors so that's a great turnout. Brian O'Neill started attending matches in the Republic many years ago and we all started to follow that. We all meet together outside Dundalk to go down to the competitions, there'd be ten or twelve lorries, all carrying tractors to the match. I have a lorry myself now and I can set off early in the morning to go anywhere in Ireland and can be home that night!

David Wright with his golden plough

The hospitality we get when we cross the border – that's what attracts us all, the way we're treated. We are now much more than just ploughmen, we are friends, we see one another in the summer and we get invited to one another's houses and the weddings.

David acts as a mentor to a number of young people hoping to get involved in ploughing. 'Our young people are our future and it's important to encourage our young people in agriculture, it's the biggest industry Ireland has.' One of David's mentees is

David Wright on his winning plot at the World Championships
in Canada, 2003

Monaghan woman Joanne Deery, who has been Queen of the Plough on four occasions.

David introduces me to the concept of the 'ploughing widow'. His own wife Rosemary says they all become ploughing widows when the season starts in September. 'I think all wives and partners would agree with that. Our heads are full of it.'

But when it comes to sponsorship, they just can't win! Potential British sponsors say they're Irish and potential Irish sponsors say they're British! We have to fund everything ourselves and that's a huge problem, especially if you're travelling somewhere like Kenya, it's hard to be competitive.'

Champion ploughman David Wright agrees, estimating the cost of transporting equipment to Kenya at up to £15,000. 'We hosted the World Championships in Limavady in 2004 and that

Five of the Wright brothers surrounded by their trophies, 1959

nearly broke us. It took a lot of money to run, we're not in debt but we haven't a lot of cash flow. There's a lot of fundraising involved.'

David has been Northern Ireland champion on four occasions and he was runner-up in the World Championships in 1998 and 2011, and won in 2003 in Canada. He has won two titles in the Republic, winning in the Conventional Class in 2002 and in the Reversible Class in 2008.

David grew up on a farm in Magherafelt in County Derry. His father Don ploughed at world level on seven occasions and his best result was a fourth place.

But David believes his father was a victim of the rule at the time that said you could only compete four times in the world event. 'He was stopped in his prime because of that, there was a huge gap until the rule was changed.'

Don was just one of six brothers, who have an enviable record. The others – Willie George, Norman, Richie, Des and Jack – all ploughed in the World Championships on a number of occasions, Des taking the world title at Horncastle in England in 1984 and in Canada in 1986. There are many accolades in between for all of them.

Then, it moved to the next generation and David started when he was thirteen years old. He competed that year in the Beginners Class in a local match in Magherafelt and he won. 'That gave me a push to keep going,' he says.

Today, he worries about the lack of young people getting involved. 'I'm forty-seven now and I'm one of the younger generation!' Other countries handle it differently. In Austria, for example, you can only compete three times and you can't plough after twenty-five years of age, 'so they have a lot of young ones as a result'.

The Wrights have been a pivotal part of the ploughing community in Northern Ireland for generations, and David says he never forgets that he owes his own success to his dad and his uncles. 'I get the credit for something my dad put the work into. They had to start from scratch whereas the skills were handed down to me over the years. They taught me everything they know.'

His mother Evelyn is the Secretary of the local association, The Loup, where William King is president. David's dad, now seventy-nine, judges at Nationals, World and European competitions and he still does the ploughing on the home farm. And David's eldest son Jack is starting to show an interest and will be competing locally later this year.

David loves coming south to compete. 'Politics doesn't come into it. It's an honour to get down there ploughing. I am a judge there too, so much competition and the enthusiasm is phenomenal. They respect the judges from Northern Ireland, we judge what we see, we don't care who's in the tractor seat.'

But, it seems, ploughwomen are in short supply in Northern Ireland, so much so that they can all name the two women who compete regularly – Pauline Davison from the North who ploughs in the Vintage Class and Joanne Deery who comes up from Monaghan. Apparently, the competition for women never took off in the North the way the Farmerette Class did in the South. 'They have a wee class of their own,' Brian O'Neill tells me, 'but they're a big attraction, people come to see the girls ploughing!'

The six Wright brothers and their father, 1962

'I'm seventy-eight now and I think I'll pack it in shortly! I think I'll stick to the judging from now on!' Brian is the president of the Moy Ploughing Society and has been ploughing for the past forty years. In 1976, he got involved in vintage tractors and bought an old American plough. He won a number of local matches with it and, in 1988, he competed in the Nationals and won the NPC Visitors' Class. He won it again in 1994, 1999 and in 2003. He also won the Northern Ireland Championships in 1991 and competed in the centenary match in Killead, County Antrim on 16 October 2016, where he came first. He has ploughed in the World Championships on seven occasions and, just before we spoke, he had won the Cavan Championship on the May bank holiday Monday.

Brian's wife Janet says ploughing is 'another religion', but Brian emphasises that going all over the country to compete is 'not a holiday' as people might surmise, though it is 'a working holiday'.

He is concerned that the numbers are shrinking a bit, although his own son Kieran has started competing locally.

> There might be only ten people taking part in a local match these days. We run a lot of matches in the winter but there is a lack of young people coming up. This is because Northern Ireland is so small. I think the ploughing will die out in the North eventually, it's heading that way. We've tried a lot of things, we've run classes for young competitors, but young people just don't want to get involved.

'Young men want to play with their big toys and the electronic stuff, they don't want to go ploughing!' That's the opinion of Thomas Cochrane, one of Northern Ireland's most successful ploughmen. 'We need new blood,' he says, 'and thankfully the

Gills and the Wrights have sons coming up now, but competitive ploughing is too expensive for young lads to get into – a brand new plough costs around £15,000.' He also draws the comparison with the NPA in the Republic which he says is a much richer organisation and can afford to pay for training and support for aspiring ploughmen.

'Competitive ploughing is too expensive for young lads to get into.'

Thomas grew up on a mixed farm in Coleraine, County Derry. There were potatoes, sheep and cereal. His father's cousin, William King, also a well-known ploughman, lived on the farm beside the Cochranes and on the opposite side was another ploughman, Andy McClements, both of whom helped the young Tom as he got interested in ploughing. He trained with the Gill brothers when he was very young 'with a wee vintage-style plough'. But when he was sixteen, his father gave him a brand new plough for Christmas.

The following day he was busy practising in the back field for the New Year's Day match in Ballycastle, when events took a very strange and tragic turn. It was Boxing Day and two of his friends were fixing an exhaust on their car so that all three could go out that evening to a Young Farmers dance.

I walked back up our big yard and, halfway up the yard, something told me, *Tom, no! Don't go out. Go back up that field and practise!* Now, this is true, I don't know what it was and, since that, I have belief in things that are meant to be. My mum Colleen played bowls and she was a member of the Limavady club which is maybe ten miles away and she was over at the bowls with my younger sister Wendy that same day. For some reason, I came back into the house, put the old clothes back on again, went by the workshop and said, 'Boys I'm back up to practise. Come back to meet me at seven o'clock, I'm just not happy with the plough and I know if I go out tonight I'll be hungover tomorrow.' So they said that's grand and I left.

At half six that night, I came in and my dad was making tea and my mum and sister weren't home yet. Two police landed at the door, we knew one of them. He said there'd been a serious accident up on the mountain and my mum and Wendy had been taken to hospital, though he said they were OK. But they had been in a collision with another car with my two best friends in it, and they had both been killed outright. I would have been in that car. My mum was in hospital for three months afterwards and, since that happened, from that day, she went to every world contest with me. She knows more about ploughing now than most men put together!

Thomas is now a respected judge at events around the country but, ironically, at one time, he felt he was getting a particularly hard time from the judges. 'Sometimes you'll know more about the ploughing than they'll ever know. Sometimes you have to hold your tongue and say nothing! Sometimes it's down to an opinion.'

Thomas Cochrane getting in some practice before heading to the World Championships in New Zealand, 2010

But it was more than just an opinion when Thomas became the youngest Irishman ever to represent Northern Ireland at the World Championships, when he went to Norway in 1989 aged twenty-three. 'I was ecstatic when they gave me the third-place medal! I was on the podium and I had the bronze medal and the trophy in my hand and a board member came over and said, "Tom, they've made an error here, we have to re-present the trophy. I have to take the trophy off you." The other man was standing beside me, he hadn't a word of English and he thought he had come fourth. So I turned around to him and I presented him with his medal. But, actually, I was over the moon. It was my first time in a world match, and here I was getting fourth! That makes you fight all the harder.'

Thomas went on to take the world title in 1997 in Australia in the Conventional Class. Then he switched to reversible ploughing and won three European titles. He says he has now qualified for and competed in more world contests than any other ploughman and he keeps coming 'in the top three or four'. He was runner-up on four occasions at the Worlds, in 2004, 2008, 2010 and 2016, when he was beaten by just one point at York in England. He also took the senior title in the Reversible Class at the Nationals in 2004.

Hugh Barr

The godfather of ploughing in Northern Ireland is Hugh Barr. Now aged ninety-one, Hugh startled the world by winning three consecutive world championships in 1954, 1955 and 1956. Bearing in mind that the World Championships only started in 1953 (when Canadian James Eccles took the first title

Hugh Barr with his golden plough

and Norwegian Odd Braut was runner-up), it was an amazing achievement to take three world titles in a row!

Hugh grew up on a farm in Coleraine just after the war. 'The grass ploughing had to be done before Christmas because of the frost in the springtime.' His older brother Robert was a ploughman who won a local match in Coleraine in 1941. Then he got Hugh involved and he took part aged fourteen.

Hugh took the world title in Killarney and went on to win in Sweden in 1955 and in the UK in 1956. He laughs when I ask him what it was like to win at the World Championships. 'I didn't let the occasion run away with me!'

Hugh and his wife Kathleen had seven children and one of them, his late son John, followed him into ploughing. John also followed his father's footsteps into the World Championships,

Hugh Barr competing at the World Championships in Oxford, 1956

competing for Northern Ireland in Australia in 1982 and in Zimbabwe in 1983 where he came third.

Hugh still follows the Ploughing but he is concerned that not enough young people are getting involved 'in this country anyway. How young people can afford it, I don't know!'

Hugh showed the way but he has been followed by Desmond Wright who won in 1984 and 1986, Thomas Cochrane who won in 1997, David Wright in 2003 and David Gill in 2007. They all have their own stories to tell.

Wexford

Martin Kehoe amid the crowds at Dublin airport, 1994

'I was after trying so often to win the world title!'

It was 1994, thirty years after Charlie Keegan from County Wicklow won the World Ploughing title, and now a new Irish champion of the world – Martin Kehoe from County Wexford – was arriving at Dublin airport to a hero's welcome! Hundreds of people were clamouring to shake his hand, to be part of this moment in history.

His daughter Eleanor remembers the occasion. 'It felt like forever waiting for him to come through the arrival gates. Mam came out first and then Dad and Jackie O'Driscoll was with him. When the crowd saw him, they just gave a big cheer and we were just pushed forward. Nobody had won the World for thirty years.'

'I'd be ploughing really well and things would just go wrong on the day.'

But this was only the beginning for Martin, who represented Ireland eleven times at the World Ploughing Championships, becoming world champion on three occasions. He also won thirteen National Ploughing titles.

'It was great,' says Martin. 'I had been trying for so long and I had been unlucky with some of the plots. I'd be ploughing really well and things would just go wrong on the day. I'd get a bad plot, practice would be going well and there'd be parts of the plot that would plough lovely, other parts of it wouldn't. That was until New Zealand when I got two reasonably good plots and they ploughed very well. I won

Martin Kehoe ploughing at the Nationals, 2011

The Kehoes – (back row, l–r) Michelle, Christine, mother Karen, Eleanor (front row, l–r) Martin, Martin Jr and Willie John

the stubble on the first day but I was only fourth in the grass, but I had enough points from the carryover so the overall got me there.'

He says the fuss at Dublin airport was unreal. 'I wasn't expecting even a quarter of what was there. A lot of people came. They tell me there were more people that night in Dublin airport than for a football team coming home. It was just beyond all expectations. And the things that went on after that, through the whole country, celebrations at dinner dances and presentations. It was a whirlwind.'

It's a cold winter's evening and there's a roaring turf fire and a pot of tea and Martin and his daughter Eleanor, herself a Queen of the Plough, welcome me into their home.

Martin tells me he grew up on a farm in Youletown, Ballycullane near New Ross. His father William used both horses and tractors and Martin says he wasn't even walking when his father brought parts of the plough into the house where he worked on them. His father was one of the ploughing legends and now Martin's children have taken up the challenge, and all five – three girls and two boys – are title holders. The girls dominated the Queen of the Plough competition from 2000 to 2012: Michelle won on four occasions, Eleanor won twice and Christine also holds a title. Willie John won the Senior Conventional Class in both 2007 and 2009 and the Intermediate Conventional in 2011. Martin Jnr won the Under-28 Reversible Class in 2012.

Martin says his own earliest memory is going to the ploughing matches with his father in 1958. 'The Ballycullane local club had just started up and I was at that match, I still remember the Kilkenny men ploughing and the neighbours all practising for the local matches to compete for the county.' From then until the mid-seventies it was all grassland ploughing, there was never much stubble ploughing. It's harder to get grass sites now. 'I first won an All-Ireland Under-21 match in Enniskerry in 1965. I was only sixteen at the time.'

Not to be outdone, Eleanor points out that she drove a tractor around the farmyard from the age of twelve and started to compete when she was sixteen. I have been waiting for someone to explain clearly the difference between 'conventional ploughing' and 'reversible ploughing' and Eleanor and Martin are the first people who can spell it out without hand gestures or assuming a basic knowledge of the operation!

'You start off with an opening split,' says Eleanor, 'that's basically breaking the ground. You set up poles to keep you straight. They're the guides for the driving, and you put them in before you do the opening split at all. Then you line up your tractor. For the opening split, you line up the tractor, you turn out a light furrow first, just one sod.'

'Then you go to the other end of the field and come back and you take two furrows and then you have to wait, that's the opening split,' says Martin. 'That's judged and then you start off again. You're looking for straightness and uniformity of the furrows, not the size but the uniformity along the field. It's harder than you would imagine. You won't get as many marks

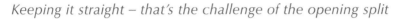

Keeping it straight – that's the challenge of the opening split

for uneven depths than if you had them dead even. They look a lot better and you'd get more points when they're even.'

'Keeping it straight, that's the challenge,' continues Eleanor. 'Vintage ploughs and horses are not so different in style because, in the earlier ploughs, the type of body on the plough was based on the horse.'

'For reversible ploughing, the equipment is different, you're carrying two bodies on the top, or five bodies,' says Martin.

'You make your crown, turn the sides back over so they meet in the middle, no gaps, no holes, no overlapping,' Eleanor interjects.

'You get half an hour for the opening split, twenty minutes in the World Championships. And two hours fifty minutes to finish your plot (two hours and forty minutes in the Worlds and it's a bigger plot). You leave only one wheel track.'

Eleanor Kehoe at the Nationals, 2011

Father and daughter finish one another's sentences, they listen attentively, while drinking tea and polishing off the Christmas leftovers.

Martin has retired from ploughing and hasn't competed for seventeen years. He went out on a high note, having won the World Championships for the third time in Pomacle in France in 1999.

Eleanor says the Queen of the Plough dance was the highlight, where the queen was actually crowned. But that's all changed and there isn't a dance anymore. The queen was usually sponsored by a shop in the town who put up a dress as a prize. 'It has become more glamorous in later years,' says Eleanor, 'but, for me, it was an opportunity to engage in a family tradition, something Dad was very passionate about. There was always a dance but that's gone for the past ten years. And there'd be a gala dance as well where the trophies were presented. All the competitors would dress up for the dance. It used to be a really big thing. Michelle was one of the last to be crowned at a dinner dance.'

Our conversation turns to what the Ploughing means today in rural Ireland. Eleanor has very definite ideas on this.

In order to win, you have to be better than anyone else. It meant a lot when my father started, there was a lot of pride then in your work, how well a farmer had ploughed his fields. The implements now, the big machine is able to till the land. A good seed bed was the essence of ploughing long ago and without having a hell of a lot of work to do afterwards. But machines are so big now and so aggressive on the soil that they're able to create a good seed bed. I feel we're honouring a tradition that represents rural Ireland, the basic fundamental of tilling the land for food. There are fewer people farming today, but the Ploughing is still a huge draw for farming people and people with roots in the soil. A lot of non-farmers like to have a vintage plough and to go out and compete today. Then there's the friendships, the social side. It's a great social life, you make friends and meet people from all over the country, you go to the local matches and have a bit of banter with everyone.

'A good seed bed was the essence of ploughing long ago.'

'You often wouldn't agree with the judges,' Eleanor continues, 'and there's often a debate over a judge's decision.' She winks at Martin, urging him to remember one 'not too far away'! But that's all they'll say about it!

We have a chat about Martin's father, William Kehoe, a champion ploughman as well, who won the National title with his horses on four occasions in 1941, 1944, 1949 and 1950 and who then went on to win the Conventional in 1955.

Eleanor reminds us that her father and John Tracey may have been ploughing rivals but they were friends as well, their children played together in Carlow. 'When we were kids we used to go up to Traceys of an evening, they were a bit older than us and they used to give the younger ones piggy backs up and down the corridors in the house!'

This type of friendship permeates the Ploughing and means that the championship itself is still seen as a community-based event, an extension of the local ploughing matches held during the year. And, with the huge number of visitors, it has become unique among such large-scale events, in that there is little or no trouble. It's just a decent crowd of people and everyone behaves.

There is also a community helpfulness in the air. I was on the receiving end of this some years ago when I lost the keys of my Volkswagen Beetle *somewhere* on the huge site. I got back to the car and didn't know what to do, I wasn't even sure that a spare key existed! Then a couple of helpful ploughmen came by, saw my predicament and set about hot-wiring the car in order to get me home! That's the sort of lateral-thinking Good Samaritan you can meet at the Ploughing.

Actually, Wexford seems to be the place to lose both keys and cars. Journalist Ray Ryan told me how, in 1998, at the Nationals in Ferns, he was driving a new car 'about a week old'.

He arrived in early in the morning while it was still dark and left the press tent in the evening, again in darkness. 'I couldn't think where I had left the car and I didn't know the new registration number. So I had to get a man with a bib from Wexford County Council with a big flashlamp. We walked and walked and walked looking for it and, after a long time, we found the car at the back of the press centre, about 20 metres away from where I'd been all day! So that was a valuable lesson, always remember where you park your car!'

The event is also a very busy time for the gardaí who have to manage a huge volume of traffic, not to mention tractors and horse trailers as well. Some of the gardaí have a special interest in the Ploughing having come from rural farms themselves. There is a Wexford garda (originally a Kerryman), who has taken a great interest in the Ploughing, retired Superintendent Peter Finn. In his book, *Plough Music: Sketches, Secrets and Stories*, the author, *Wexford People* reporter David Medcalf, tells Peter's story, starting in 1994 when the championships were held in Enniscorthy.

As a countryman, Peter approached the ploughing-match assignment with a sympathetic, empathetic, accommodating attitude. In his youth, the superintendent had himself ploughed with a pair of horses – one black and one grey – on the Finn family's 35-acre holding in the west of the Kingdom. In 1973, the then twenty-one-year-old, a raw garda recruit, was sent from the barracks in Arklow to rain-swept south Wexford. That year's Rosegarland Championships are still remembered for putting the wellington into Wellingtonbridge. He carried out his duties in a sea of mud, sustained by official rations that extended to a bottle of Taylor Keith and a ham sandwich, an experience that led to him setting up a decent Garda catering unit so at least the officers under his command could expect better fare.

My journey among the ploughing people took me to towns and small villages where the heroes of the past are still spoken about, and where the polish of a furrow and the shine of a harness are still admired. For these families, ploughing is still a way of life rather than a remembered tradition. And there is huge importance in what these families are doing today because, if we're not careful, it could all disappear within a generation – and what a loss that would be!

As a child I spent holidays in a cottage in Lower Rosses Point in County Sligo, and the impact of this rural idyll lingers in my memory. We went with my dad to gather mushrooms with the dew still on the grass. We ran wild through the fields and across the strand. We crossed the stream at low tide to reach Lissadell, and shivered as lightning shook the slopes of Ben Bulben. We spent days in the fields as the hay was saved and made into giant stacks. During those days, the farmer's wife would bring mugs of tea and plates of brack to us, while we listened to the men as they told their stories of tying their trouser legs so that the rats in the hay wouldn't bite them.

And, of course, the past was a sunny place! But these moments in time from my memory are exactly what our ploughing people continue to create. Yes, they till the soil to grow food but they are also continuing a way of life that casts memories and shadows over the people and the landscape of rural Ireland.

5. The Farmerettes

'The word is ridiculous! They're all farmers! A farmerette? Anything that's an *ette* is always smaller, not quite the full thing. I have never agreed with it, they could say "Mixed Classes" or "Ladies' Classes" but what's wrong with the word "woman"? We don't say "Gentlemen's Classes" – it's totally outdated!' That's what Mairead Lavery, editor of the *Irish Farmers Journal Country Living* magazine, the biggest selling women's magazine in the country, thinks of the term *farmerette*.

The farmerettes are part of J.J. Bergin's legacy. The idea being that the winner would become Queen of the Plough. There was a lot of opposition to the idea, but he got his way in the end, and it was decided that the competitors could be single, married or widowed. Larry Sheedy described the scheme as 'the promotional coup of the decade'. 'It was the prototype of

'What's wrong with the word "woman"? We don't say "Gentlemen's Classes" – it's totally outdated!'

promotional strategy copied and used by a new generation of young industrial giants in the next decades. The Queen of the Plough was literally seized and feted.'

But is it a suitable title for a competent plough person who happens to be a woman? Sheedy describes how, in 1993, the *Irish Independent* thought it would make a good story, and so the story grew legs, with the Council for the Status of Women demanding that the name be changed. 'But,' says Sheedy, 'the NPA was on cast-iron ground, with a woman as managing director, and a squad of women who were entirely proud to compete as farmerettes, regardless of what exactly it means.'

But let's turn the clock back to 1955 and the first farmerettes. The first Queen of the Plough was Anna Mai Donegan from County Kerry. Then, in 1956, a further incentive was added: if the Queen of the Plough married during her reign, she would receive £100 on her wedding day. While Anna Mai reclaimed

Anna Mai Donegan being crowned Queen of the Plough by John Tracey, 1956

her title, the cheque remained unclaimed, as it was a few years before she married!

Women had always worked on the farm, but it was really the sharp eye of the tractor salesmen for promoting their products that enticed women into competitive ploughing. In Anna Mai's case, it was Martin Slattery, the salesman from Benner's Garage in Tralee, who came up with the idea. Interestingly, Mr Slattery was mentioned to me several times as a go-getter tractor salesman who saw the advantage of attractive young farmerettes being photographed on his tractors!

Anna Mai had always been driving a tractor round the family farm. 'Time has changed everything. There's lots of women on the farm now driving tractors. But you have to have it in you to master driving a tractor, back then very few women did. When farmers saw a woman ploughing, they were all surprised and they'd call in to see how you were doing. At mass, they'd tell me I was wonderful to be driving the tractor.'

One day Martin Slattery came to see her and asked if she'd try the Ploughing.

'When farmers saw a woman ploughing, they were all surprised and they'd call in to see how you were doing. At mass, they'd tell me I was wonderful to be driving the tractor.'

I laughed at him. I grew up on a 30-acre farm with animals, tillage, everything. He said he'd seen me on the tractor and eventually I agreed to compete. Martin became my coach and, in 1955, drove me to the first championships in Athy, County Kildare, in a Ford car. And I won! I was thrilled of course, although I don't remember actually getting a prize for it. I was very excited and dying to tell my parents my great news. They really encouraged me but there were no telephones and I had to wait till I got home to tell them I'd won.

Anna Mai Donegan on her family's farm

Anna Mai retained her title the following year and entered the Nationals again in 1957 in Nenagh, County Tipperary, but she exited the championships after she was given a bad plot of ground. 'It was water-logged, you couldn't plough it, so I pulled out of the competition and I never went back!'

Anna Mai says her parents enjoyed the Ploughing and really encouraged her, and it's only in recent years that she realises how big a step it was for a woman back in 1955. 'I was good at keeping the furrow straight, I was mostly self-taught, a lot of practice. I did it for fun when I was asked originally because I was always out and about on the tractor.'

And there was another treat in store as well, a trip to an agricultural show in the UK, chaperoned by the NPA's Larry Sheedy. Larry remembers it well. 'The winner would get a trip to England for the Smithfield Show. The first time we had that, in 1955, Anna Mai Donegan was the winner and I had to go with her to keep an eye on her!'

'It's still a great place to meet a farmer.'

Anna Mai married a farmer, Robert Gleasure from Clogherbrien, and the pair regularly travelled to the Ploughing Championships. 'It was a big trek in the early years, we used stay for the three days. We'd meet up with lovely people and there'd be parties and dances every night. It's still a great place to meet a farmer.'

Next I head to Abbeydorney in County Kerry, which has had an inordinate share of winning ploughmen and women. I'm here to talk with Mary Shanahan, but first I have to contend with her husband, Monty, who also has an opinion on the farmerettes. 'They were well known for one reason – for the young fellas following them around the field.'

Mary Shanahan ploughing in the Championships, 1959

'That's jealousy!' retorts Mary. 'I wasn't conscious of lots of young men following me around.' However, under scrutiny, we get a unique insight into the world of the farmerette. Yes, they get asked out by lots of young men 'but I didn't take any notice – they all lived too far away, there was no point'.

But maybe there *is* something in the air. I ask Monty if she was fussy.

'She wasn't fussy at all. I'll tell you what she was, she was very shy! Am I right, Mary?'

At this point, the two romantics forget about me and head off into how they met one another at a film in the local hall.

'There was a man worked in the creamery and they'd given him the bicycle to go home, so he took the bike and I'd gone over to collect it. And there was a film on in the hall. Monty was sitting in the hall beside me …'

'You came and sat down beside me!'

'So, I said, "Any chance you would drive me home?"'

'You were trying to avoid your man at the door,' says Monty, 'and I said, "What's wrong with you?" And you told me he was tormenting you to carry you home on the crossbar. And I said I'll throw up the bike on the back of the car for you.'

The couple never got engaged but got married four years later.

Mary was the Queen of the Plough in both 1959 in Kilkenny and 1960 in Wexford, when she had a great year, the highlight of which was driving her tractor at the head of the St Patrick's Day Parade in Dublin. She got great cheers all along the route, plenty to remember for the future.

We talk about the future of the Ploughing and Monty has plenty to say about that too. 'Now they have no virgin ground left to plough, the only thing they have now are cornfields and it

Mary Shanahan leading the St Patrick's Day Parade in Dublin, 1961

beggars belief how they can judge it because, when you turn it over, half an hour later it all looks the same.'

Mary agrees. 'All you see is stubble, that's all you'll see, sticking up. It's all rough, there's no sleekie, no polish, like my father used to have.'

Since that first competition in 1955, there have been sixty-two Queens of the Plough from counties Kerry, Wicklow, Galway, Kilkenny, Laois, Waterford, Wexford, Dublin, Cork, Carlow, Offaly and Monaghan. Many of them hold several titles; Breda Brennan-Murphy from Carlow holds the record, having won the event on five occasions in 1994, 1995, 1998, 1999 and 2004. Joanne Deery from County Monaghan is a

Joanne Deery, four-time Queen of the Plough

close second with four titles in 2011, 2013, 2014 and 2015. Thirty-seven-year-old Joanne comes from Iniskeen, from good ploughing stock. Her grandfather, John Deery, competed in the horse ploughing at national level, coming second. Her dad, also John, a dairy farmer, was placed second in the Junior Class in 2016 and his brother came third. She also had a role model in her mother, who ploughed in local matches.

Joanne herself started competitive ploughing when she was sixteen and her younger brother has been third and fourth in the Under-21 Class at the Nationals. In 2017, another generation enters the frame when her eighteen-year-old son, Aidan, enters the Novice Horse Ploughman event. Joanne is very excited about this because 'we're trying to promote horse ploughing in Monaghan, it's nice to see the horses continuing'. Aidan is being trained by Noel Hand, who is also a competitive ploughman.

Joanne says she is lucky to have the former world champion David Gill as her coach. They get on well together because of 'David's patience and my interest'. Her first foray into the Nationals was in 1996 at Oak Park in Carlow. But 2011 was her first big win. 'It was a bit surreal,' she says, 'because I knew we'd done well, but it's a team effort. Dad had set everything up for me, got the tractor and the equipment down to Carlow. You don't really win it for yourself, you win for the team! I was fifteen years ploughing before I won!' The only occasion when she didn't participate was in 2007 'when I was pregnant with one of my three boys'.

But 2011 was exciting. 'We had done up the plough and invested a good bit in it. Dad had taken a break from ploughing for about ten years, then he came back and started ploughing again himself. It was just one of those things, it all fell into place. I couldn't believe it, it would have been very close, Anna Marie McHugh was second that year and I had a feeling it was either me or her. Dad was delighted, you'd think he had won it himself!' In the Farmerette Class, the rules allow for a helper on the headland for instruction and adjusting the plough but the helper is not allowed on the actual plot.

I asked Joanne what she thinks of the term *farmerette* to describe the class. 'Well, it has to be called something. I plough in Northern Ireland and up there it's known as the Ladies' Class, really it makes no difference but it is a very historic class, there's a huge number of big names in it.'

Joanne says the counties around the border are quite lucky. 'Living further south, people think of the North and South as two different entities. But we're on the border and we wouldn't have our local matches without them, they would support all our local matches around Meath, Cavan, Longford, Donegal and

Monaghan. We would see them every week in the springtime.'

Ploughing really is a way of life for the Deerys. Apart from the plough people, Joanne's sister Róisín works for the week of the ploughing at NPA headquarters and has also assisted at the World Championships. Another sister, Áine, ploughs at novice local matches. 'She's getting married in June, she wouldn't be allowed get married in September – that's reserved for the ploughing! Áine and her mum are chief caterers at the local matches around Monaghan. Áine is also a nurse and provides first aid at the matches.

'She's getting married in June, she wouldn't be allowed get married in September – that's reserved for the ploughing!'

So, what about the future? 'Ploughing will survive,' says Joanne, 'in every county there is a stalwart, somebody who will keep it going, it will be there for a good while yet. Also, there are so many family connections, people will always look out for such and such a person's son, daughter or grandchild.'

Being a farmerette has always been seen as a bit glamorous, even if there's boilersuits and wellies involved, but everyone who talks about the farmerettes also emphasises their sheer skill, strength and dogged determination.

A farmer at the Roundwood Ploughing Match in Wicklow said, 'They're doing as good a job as any man!'

But not everyone thinks like this. In September 2012, in a letter to *The Irish Times*, Arthur Brady from Dublin 14 wrote, 'Sir, I see that the *farmerettes* are alive and well at the Ploughing Championships in New Ross. It's enough to put me back on the female cigars! – Yours, etc.'

Mairead Lavery has been very involved with women in farming through her work for the *Irish Farmers Journal* and she says women have proved they can do any function in agriculture. 'They're highly capable, organised, skilled, they can adapt to new processes very quickly. I really believe that farming cannot survive without women.'

But still the number of women running their own farms is very small as the tendency still is for it to be inherited through the male line. In Ireland only about 12 per cent of farms are owned by women compared with 33 per cent in Austria. Mairead says, 'They are mostly widows who have inherited the farm. But there is nothing to stop women progressing, the courses are there, the role models are there.'

Mairead Lavery, editor of the Irish Farmers Journal Country Living *magazine*

Mairead's own story from the Ploughing comes from 1991 when she dressed her four-year-old twins in 'little red wellies and little blue raincoats – they were just gorgeous and everyone was admiring them'. It was a very mucky championships and, at the end of the day, she went to a tap to rinse their wellies. 'They stepped into the puddle beside the tap and they were up to their waists in water! They disappeared in front of me! They were absolutely saturated. They've never forgotten it even though that was twenty-five years ago. We had to strip them off, get them into the car and wrap them in rugs!'

6. The Venues

Seven hundred acres, Portaloos for 300,000 people, 24 kilometres of steel trackway and medical emergency rooms. This is part of the shopping list for P.J. Lynam, the outgoing Chairman of the NPA, as he gets yet another site ready for the Ploughing – the latest is Screggan in Tullamore, which also hosted the event in 2016.

Work starts early in the site. 'We start working for the next year on the day after the competition ends,' he says. 'And there's heaps of work to be done even at that early stage, we have to map out where the ploughing will happen. This year, in Tullamore, there will be seventeen different farmers involved. Joe Grogan is the main farmer and then all the others will let us use their land. The infrastructure is a huge task. It's the very same as providing a fully serviced building site, including everything down to the internet.'

I wondered if anyone had chosen the Ploughing as a wedding

venue yet. 'There have been no weddings yet but we had a church, a white church at Ratheniska in 2015,' said P.J. 'They rent out the church so you can have a wedding in your own garden. I'm sure we'll have a wedding here yet!'

P.J. Lynam, outgoing Chairman of the NPA

And they've found a way to combat the problems of getting people and vehicles through the muck and the mud by importing steel trackway from Germany and the UK. 'Nobody does it here so we have 24 kilometres this year and that starts going in in the last week of July.'

The full-time work on the site begins officially on the first Monday in August. When the toilets, all Portaloos, are put in and the water and electricity are installed.

'It's the place everybody wants to be in September.'

'We have to have everything ready for around 100,000 competitors and visitors a day. There are also medical emergency rooms, last year we had a few falls and a couple of heart attacks at the show. We are very fussy about health and safety, this is not a fly-by-night event, it can stand up to any scrutiny. It comes down to this,' he says, 'the Ploughing is at the heart of rural Ireland, and it's the place everybody wants to be in September. The harvest is finished and it's a kind of a holiday for the farmers.'

The site at Screggan before the 2016 Ploughing Championships

Around 24 kilometres of steel trackway is used to keep the mud at bay

And the commercial interests are determined to be there too; trade bookings for 2017 started coming in the same week the 2016 show ended. There are strict regulations for stand holders and everyone has to agree to co-operate. 'If anyone creates a problem then they're not allowed back,' he says. 'For example, if someone takes a 12 x 4 stand, but they start selling things outside that, then it's goodbye.'

P.J. doesn't think a permanent site would be a good idea. 'No, people enjoy the change around, the different venues.'

But some people I meet are not so sure about that. They remember the years when there were severe weather problems – they all mention 1974 and the World Championships in Wellingtonbridge in County Wexford when there were gale-force winds strong enough to take the tents down.

They also remember the 1965 championships in Enniskerry, County Wicklow, where every bit of bad weather made an appearance and, for the first and only time, the championships had to be postponed for a week. Not to mention Kilkenny in 1986 when a swarm of midges descended on Woodsgift and attacked competitors and visitors alike. But the man who probably suffered most was the legendary Diceman, Thom McGinty, on his first visit to the Ploughing, who was covered in tasty greasepaint. Perfect for attracting midges!

The choice of venue has always been one of bringing the Ploughing to the masses and some championships have been held close to urban areas to encourage non-farming folk to attend. Indeed, some of them catch the ploughing bug and take up ploughing and even loy digging as a hobby. The event has been held in Dublin four times, in Clondalkin (1933), Cloghran (1942), Balbriggan (1946) and Finglas (1971).

There has sometimes been controversy about the choice of

venue. In 1984, for example, Ardfert in County Kerry was chosen, but some people thought the event was too big for such a small village. However, the NPA insisted on their policy of bringing the Ploughing to the people. The Kerry organisers were adamant it was a good idea, and one of them was quoted saying, 'The critics thought we had no horses in Kerry for the visiting ploughmen, but we turned out reams of big strong horses that shook the main street in Killarney when they trotted through.'

'We turned out reams of big strong horses that shook the main street in Killarney when they trotted through.'

The *Irish Farmers Journal* reported that the championships were an astounding success.

> The most amazing feature of this year's event was that, despite the remoteness of the site in relation to the main tillage areas of the country, there was a capacity crowd present. It surpassed all expectations, particularly within the farm machinery grade.

Marking out the plots in Danesfort, County Kilkenny, 1970

The *Journal* and, notably, reporter Noel O'Reilly have commented every year on the choice of location, the suitability of the soil, the impact of the weather and the skill of the ploughmen. He wrote of the 1971 event:

> Disappointing attendances at last week's National Ploughing at Finglas brings home very forcibly that location is all important for an event like this. The NPA have learned their lesson the hard way and are not likely to be caught out again. Gate receipts for the two days were down by £1,200 and this has always happened any time the ploughing match has moved out of the traditional tillage areas of the midlands and south. On the surface, this year's site appeared to have everything. Besides it was only a few miles from Rush, one of the most intensive tillage areas in the country. The difference is that there is little real interest in ploughing in the area and, apparently, a day out does not have the same appeal as it does in other areas like Kilkenny, Carlow, Wexford or Tipperary.

But there was praise for the New Ross venue in 1960, the *Journal* calling it 'ideal'. 'The competition fields were conveniently located and compact, and trade stands were positioned so that the crowd watching the ploughing circulated around them on both days.' Conditions on the first day were 'delightful' according to the paper but then came the rain, 'making things very difficult for competitors'.

Larry Sheedy remembers the early years when 'there were hardly any spectators, just a couple of caravans and maybe a bottle of whiskey in the members' caravan'.

> Back then, it took place in February in the most god-awful weather you could possibly get. They ploughed in frost and

The Ploughing Championships in Killarney, County Kerry, 1939

snow and then as time went on and people started to harvest early, they could harvest in August. So, when that happened, they started to hold the championships in September.

Larry's own bad-weather story comes from 1987 in Tullamore when there was the most dreadful conditions. 'Nobody took the Ploughing too seriously, including the gardaí in Tullamore, they didn't even switch off the traffic lights for the traffic that was coming. They had ferocious traffic jams and then, when you went to the field where they were parking the cars, it was flooded so you couldn't drive your car in and park, it was just a mess. It created a lot of muck and a lot of anger with people who couldn't get in and do their business. I was doing PR for them at the time and I wrote a memo

'They had ferocious traffic jams and then, when you went to the field where they were parking the cars, it was flooded so you couldn't drive your car in and park.'

Venues of the National Ploughing Championships

1931 Athy, County Kildare
1932 Gorey, County Wexford
1933 Clondalkin, Dublin
1934 Athenry, County Galway
1935 Mallow, County Cork
1936 Tullamore, County Offaly
1937 Greystones, County Wicklow
1938 Oak Park, County Carlow
1939 Killarney, County Kerry
1940 Thurles, County Tipperary
1941 Cork, County Cork, and Navan, County Meath
1942 Cloghran, Dublin
1943 Portlaoise, County Laois
1944 Ballinasloe, County Galway
1945 Tipperary Town, County Tipperary
1946 Balbriggan, Dublin
1947 Maynooth, County Kildare
1948 Limerick City, County Limerick
1949 Drogheda, County Louth
1950 Bandon, County Cork
1951 Wexford Town, County Wexford
1952 Athenry, County Galway
1953 Mullingar, County Westmeath
1954 Cahir, County Tipperary
1955 Athy, County Kildare
1956 Nenagh, County Tipperary
1957 Boyle, County Roscommon
1958 Tramore, County Waterford
1959 Burnchurch, County Kilkenny

1960 New Ross, County Wexford
1961 Killarney, County Kerry
1962 Thurles, County Tipperary
1963 Athenry, County Galway
1964 Danesfort, County Kilkenny
1965 Enniskerry, County Wicklow
1966 Wellingtonbridge, County Wexford
1967 Tullow, County Carlow
1968 Banteer, County Cork
1969 Cashel, County Tipperary
1970 Danesfort, County Kilkenny
1971 Finglas, County Dublin
1972 Cashel, County Tipperary
1973 Wellingtonbridge, County Wexford
1974 Watergrasshill, County Cork
1975 Bennettsbridge, County Kilkenny
1976 Gorey, County Wexford
1977 Cashel, County Tipperary
1978 Knocktopher, County Kilkenny
1979 Watergrasshill, County Cork
1980 Cashel, County Tipperary
1981 Wellingtonbridge, County Wexford
1982 Edenderry, County Offaly
1983 Waterford City, County Waterford
1984 Ardfert, County Kerry
1985 Athy, County Kildare
1986 Urlingford, County Kilkenny
1987 Tullamore, County Offaly
1988 Oak Park, County Carlow

1989 Oak Park, County Carlow
1990 Oak Park, County Carlow
1991 Crecora, County Limerick
1992 Carrigtwohill, County Cork
1993 Clerihan, County Tipperary
1994 Enniscorthy, County Wexford
1995 Ballacolla, County Laois
1996 Oak Park, County Carlow
1997 Fiveally, County Offaly
1998 Ballycarney, County Wexford
1999 Castletownroche, County Cork
2000 Ballacolla, County Laois
2001 Ballacolla, County Laois
2002 Ballacolla, County Laois
2003 Ballinabrackey, County Meath
2004 Athy, County Kildare
2005 Mogeely, County Cork
2006 Grangeford, County Carlow
2007 Tullamore, County Offaly
2008 Cuffesgrange, County Kilkenny
2009 Athy, County Kildare
2010 Athy, County Kildare
2011 Athy, County Kildare
2012 New Ross, County Wexford
2013 Ratheniska, County Laois
2014 Ratheniska, County Laois
2015 Ratheniska, County Laois
2016 Screggan, County Offaly
2017 Screggan, County Offaly

of ten things they should do in the future. Anna May McHugh and the council approved it. The first suggestion was that they should make it a three-day event and they should hire an expert on traffic management. The next year, the three days made a huge difference and the expert traffic management created free flow.'

A central location pleases a lot of people in terms of access and transport. Athy in County Kildare hosted the original championships in 1931, and also hosted the silver jubilee ploughing match in 1955 and the eightieth anniversary in 2011, as well as the 2004, 2009 and 2010 championships.

Oak Park in Carlow has had more championships than anywhere else, being hosts on six occasions and the venue for the World Championships in 1996.

In 2013, 2014 and 2015, the ploughing went to Ratheniska in County Laois. One man who knows the venue very well is RTÉ's Sean O'Rourke as he grew up literally beside the field. So, broadcasting the *Today* show from the ploughing has been, in a sense, like going back to his roots to bring not only the sounds but sometimes almost the taste and smell of the event into homes around Ireland.

Father of Irish ploughing

ON Thursday and Friday next, the 25th National Ploughing Championship, organised by Mr. J. J. Bergin, will be held, and no doubt next week's Championship will be successful as every Championship Match that has been held since 1931.

To-day when there is a little prosperity in farming the difficulty in organising a National Ploughing Championship is not very great, but during the early 30's when the economic war was at its worst and farming carried few compensations, the organisation of a Ploughing Champion-

J. J. BERGIN
... in good times and bad ...

ship must have been no sinecure.

In those days Mr. Bergin worked hard for the National Ploughing Association and to a large extent it is due to his efforts that to-day it is an accepted part of national agricultural activities.

Coverage of the 1955 National Ploughing Championships, in the Irish Farmers Journal

Catering for 275,000 people in Ratheniska, 2015

DID YOU KNOW? With over 1,400 exhibitors, there is a lot of eating and drinking done at the Ploughing. You might not believe this, but during the 2015 event:

80,000 cups of tea and coffee were drunk each day

40,000 breakfast or breakfast rolls were sold each day

16 tonnes of prime Irish beef is consumed, as well as 3 t. of locally sourced pork

5 tonnes of cheese went into burgers and sandwiches

5 tonnes of Irish-sourced soft fruit and veg was used

1 chicken fillet was sold every ten seconds

3,500 boxes of fresh Irish salads were produced

24,000 litres of milk were consumed

Approximately 14 acres of Irish potatoes were grown to supply the event

1 Irish-produced quarter pounder burger was sold every second

Sean's father was the master of the local national school. 'It was a two-teacher school, himself and Mrs McCormack, she taught the junior classes up to first class and then he taught the rest of them up to seventh. It was a real idyll then, Ratheniska consisted of three roads meeting at a church and a hall, and the school was down the road, and there was a house going with the teacher's job and we lived there.'

Sean was only five and a half when the family moved from Ratheniska but says that even at that young age he was aware of the whole idea of the Ploughing in the area. 'One of my earliest memories was hearing how a man called Tom Fingleton, one of my father's past pupils, had won a national prize at the Ploughing. That would have been just after or just before we went to Galway. It was a big deal, the Ploughing was just so intrinsic to the place. Maybe it is an exaggeration to say it was as important as the GAA, but it was part of what they were. So I was always aware of it and I never had the opportunity, work-wise, to be part of the thing until I got the best job in Irish broadcasting, in 2013!'

Sean has presented four shows from the championships.

'It was
a big deal, the
Ploughing was
just so intrinsic
to the place.'

The extraordinary thing was that the very first that I did was right back over the hedge from the house that I spent the first five years of my life in! It was literally over the fence, over the ditch from the back garden into Carters' farm or field, and what happened in the intervening half century from when we left Laois to go to Galway, fields had been merged and it had become a vast prairie, big enough to take the Ploughing. And all those people – the Fingletons and Carters and people like that – they would have been the neighbours. And it wasn't as if any of us hadn't been back, we always went back for summer holidays, but you have to remember Ireland wasn't prosperous until the nineties really, and the idea of going anywhere on a holiday other than somewhere with a relation or a friend just didn't exist until I was in my twenties. So we used to go back to that Ratheniska area, mainly to the Fingletons, to do farm work, fork silage, work the harvest, feed the pigs, learn to milk the cows, slowly.

Sean remembers when he was first asked to take the show to the Ploughing. He was thrilled. 'It was kind of an emotional thing as well because you felt you were kind of going home, and it was in the early days of doing the show, I think I was only three weeks in and the jury was still out! At the first show, I was asked to open a historical ploughing exhibition for County Laois which was a really nice thing, and who was there, a friendly face in the crowd, but the aforementioned Tom Fingleton – eighty years of age then, still going strong, and his eldest son, Kevin.'

'It was kind of an emotional thing as well because you felt you were kind of going home.'

Sean and his parents were very impressed by the Fingletons' work ethic, 'very hard working farmers and very progressive'.

My parents remembered when they came to mass on a Sunday morning, in the chapel, in a horse and trap like all the neighbours did, and then they sort of graduated on to cars. The Fingletons had the best tractors, I remember actually going on holidays to them. They had what looked to my thirteen-year-old eyes like a huge combine harvester, a Claas, and this was about 1968, and it's the kind of thing that would maybe cost a contractor €300,000 now. And they had their own one, and Tom used to drive it. They were serious tillage farmers in the sixties and seventies and then it went down the generations, like his son, Ciaran, took over their farm, but they ran a kind of collective, like the three lads would work on one farm one day, they'd be down in Stradbally another day and they would have had other lads working there as well.

Sean O'Rourke greets Marty Morrissey as he broadcasts from the Nationals at Screggan, 2016

Sean loves the buzz of an outside broadcast. 'They're hard work for the reporters and producers, I just kind of swan in and pick up the brief and you meet the listeners and all of Ireland is there.' He singles out the queue to meet the hurley makers, the Cannings. 'The queue is probably about twenty minutes from when you go into their tent.' And one of Sean's brothers,

Caoimhín, a Jesuit priest, has also contributed to the Ploughing with a 'Thought for the Day' on a ploughing theme to coincide with the event. He says the Jesuits went a step further, setting up a tent at the Ploughing – and, indeed, there is now a range of religious possibilities for visitors.

Sean says the atmosphere at the Ploughing conjures up all that is good in rural Ireland.

> You get a sense of industry, excitement, satisfaction at a job well done at the end of the harvest season, the agricultural year. Now we were fortunate – four years I've done it, four fine weeks, I think there was a wet day as I was leaving in 2016. But you got a real sense of a country at ease with itself, rural meeting urban. I wondered afterwards if they would be better off having a fourth day – it was almost uncomfortable moving around some of the time, you can't walk. It's like Shop Street in Galway on Christmas Eve, for three solid days. You arrive at the crack of dawn to set up your stuff, and the tailback of lights from traffic on the way in. It's incredibly well organised! I stayed locally with Brendan Fingleton, when it was in Ratheniska and it was just fantastic because you'd have the bed there, and you'd know the back road down and then you join the traffic. But people must be getting up at midnight to get in.

'But you got a real sense of a country at ease with itself, rural meeting urban.'

And, as everyone tells me, it's definitely the place to meet people. 'A long-lost cousin of mine who is a sheep farmer from West Kerry, he sticks his head in and says hello!'

Sean also comments on the 'wholesomeness' of the event. 'You do not see rowdiness. I'm sure people enjoy a few drinks but people are well behaved at it, there are all ages there.'

National Ploughing Championships, Tullamore, County Offaly, 2007

And would he go along to the Ploughing if he didn't have to work at it? 'Absolutely! And I will when I retire!'

Accommodation at the ploughing is the stuff of myth. There are stories of houses being rented to four or six people and then all the cousins and neighbours arrive with nowhere to stay and they cram in another twelve. P.J. Lynam says he once saw eighteen people staying in a caravan in Cashel but that's all he's saying – 'What goes on tour stays on tour!'

The Ploughing means a bonanza for local householders, some of whom go home to mammy and daddy for the few days and rent out their homes for around €800. There's a story told of one young woman who rented out her house in Tullamore in 2016 and then went into the local hotel and used the money to put a deposit on her wedding. But, for some, there is no accommodation and, as we've heard elsewhere in this book, many a ploughman has spent the night in a horse box.

7. Running the NPA today

Denis Keohane

'**G**uiding a multimillion-euro institution requires a sound head and great organisational experience, and Denis Keohane is a man with all these requirements, and always lets the head rule the heart.' That was the reaction of John Sexton, journalist, author and Timoleague Ploughing Association member, writing in the *Southern Star* on the election at the end of April 2017 of Denis Keohane as Chairman of the Executive Committee of the National Ploughing Association. The last Cork man to hold this position was John's father, Larry, back in 1961.

Denis comes from Ballinascarthy in West Cork and, on his election, the members of the Clonakilty Ploughing Association, of which he is a member, gave him a great welcome home. According to the *Southern Star*, he was given 'a rousing welcome with a parade through the town, followed with a reception in the Brewery Bar. In an address of welcome, Gordon Jennings, who presided, said that it was a proud moment for the Cork West Ploughing Association to have one of their members bestowed with the highest accolade of the National Ploughing Association.'

'A very broadminded man who would always put the interest of the common good before everything else', was how Vincent Beechinor described Denis.

Also at the AGM, Padraig Nolan was elected as vice-chairman, the first officer of the NPA from County Roscommon.

I caught up with Denis when the celebrations had died down and he told me about his roots in ploughing and how, while still very young, he had spent seven or eight years ploughing the family farm with his father. His own son is presently studying for his Green Cert to take over running the farm and continue the family tradition.

But, apart from three or four local matches, Denis pursued the role of organiser rather than ploughman. In the sixties, he became a member of his local Macra in Ballinascarthy. Anyone who comes from the area between Bandon and Clonakilty will inform you that this is where Henry Ford came from, and on cue Denis pops it into the conversation!

The first match in Ballinascarthy was in 1969 and this is how it came about. Denis was at a Macra meeting one night when another lad asked him to propose holding a ploughing match and said he would second the motion. He did and they were left to organise the first Ballinascarthy ploughing match.

Over the past fifty years, Denis has been climbing the ranks in the world of ploughing. When the ploughing left Macra, Denis became secretary of the Carbury Association for two

Denis Keohane, Chairman of the Executive Committee of the NPA

years, then he became chairman and, in 1980, he got married. 'I stepped back for a while after that,' he says.

He is credited with the big decision in 1984 to divide Cork into East and West for ploughing competitions, giving it two-county status. 'They were travelling too much, it's a huge county and it meant that people going to local matches might have a round trip of 200 miles ahead of them. It was too far.'

Denis has great respect for the champion ploughmen who represented their county. The names trip off his tongue, people like Jerry Horgan, Cyril Dineen, the two Keohanes, Thady Kelleher, J.J. Delaney and John Sheehan to name just a few of the big achievers.

'People going to local matches might have a round trip of 200 miles ahead of them.'

Denis became Director of West Cork at the NPA and has been a member of the council for over twenty years. He has been vice-chairman for the past three years. He is a judge at national and international competitions and now, he says, he's looking forward to being chairman for the next three years, 'if my health spares me'.

He tells me a story of a newspaper discovered ten years ago when the flooring on an old pub in Clonakilty was being replaced. The men working on it found a copy of the local paper, the *Skibbereen Star* dated 1905. It carried a report of a local ploughing match that year. 'This was even before the National Ploughing got going, it was the result of a match held on the edge of the town. The land is built on since, there are houses there now. There had been earlier matches but this was the first time we'd seen a report of one so long ago. Two of us, myself and the Secretary of the Clonakilty Ploughing, went into the offices of the newspaper and we spent the day going through the old papers. Of course it was all horse ploughing back then.'

And what about the future of the Ploughing?

'I think it will survive. It's got so big, but I can't see the number of ploughmen getting any bigger.'

Denis says that enticing young people to get involved has been very difficult even with training and grants. It's one of the areas he wants to keep working on but he accepts that ploughing is a major commitment for young people to take on. He says the NPA has been very good at encouraging young people to join up. He has very definite ideas about the sort of chairman he wants to be and, of course, his track record with the NPA goes back more than fifty years, with more than twenty spent on the National Council.

I hope I'll still be talking to as many people as I am now, because I think that would be one of the things I do, I converse with a lot of people, that's what I'd like to do. And I like to go to the ploughing match and not meet the people running the ploughing but talk to the people coming along to the match. I think I get on with the ploughing people better than the GAA fellas!

And, inevitably, the talk turns to Anna May McHugh.

'Anna May, she has to be – and there's no question at all about it – the leading person in the country. In my time, I have never dealt with anybody like her. Her foresight is unbelievable – no matter how big or small it is, if somebody even throws an idea at her, if she thinks that's right for the ploughing, she will go with it and put her all behind it. She thinks of a load of things and leads from the front but if somebody comes up with a good idea, she will back it 100 per cent – it makes no difference whether she thought of it or not. For that reason, I think she is unique and she can bring people together, bring people with her, bring people around her. She knows how to manage people. I'm looking forward to working with her.'

'If somebody even throws an idea at her, if she thinks that's right for the ploughing, she will go with it and put her all behind it.'

Anna May McHugh

Anna May McHugh took over as Managing Director of the NPA in 1973 after the death of Sean O'Farrell. Now in her eighties, Anna May is content to keep working. In 2013, she was a guest on *The Late Late Show* and Ryan Tubridy asked her, 'Does the word *retirement* mean anything to you?'

'That thought has never crossed my mind!' she replied. 'It's not the years in your life that matter, it's the life in your years!'

Matt Dempsey of the *Irish Farmers Journal* has assessed Anna May's great strengths:

It's her relationship with the landowners, the host farmers. Suddenly, the President comes and your family are seated beside him, and treated as a genuine VIP for the three days of the event. It's a tremendous ego trip, it's sort of a measure of acceptance of the family's standing in the community that their family has been chosen as the centrepiece.

And that standing goes beyond the community involved, to the parish, the towns and villages, and even the county, instilling a sense of pride and belonging in anyone with even the slightest connection with the Ploughing.

Matt also credits Anna May for not making the same mistake as the Royal Show in England. 'Letting themselves be taken over by the equestrian and hobby-type things. She has kept it hardcore in visibility terms, but that hasn't stopped her expanding the number of stands.'

Anna May was only seventeen when she joined the NPA.

She told Ryan Tubridy how it came about 'by accident more than design'. J.J. Bergin was friendly with her father and it was coming up to the annual Championships in Wexford in 1951.

> I was doing a commercial course and he asked my father if I would give some assistance in the office, which I did. I went down, and all I could see was this very old, grey-haired man with a big moustache, and I spent two days there and I said, 'I am not going back there,' and that was for definite! So Monday came along and I did go and, believe it or not, I never left the job.

But she still hadn't crashed through the glass ceiling. 'The phone would ring in the office and they would want to speak to Mr Brennan because Brennan was my maiden name and I'd have some convincing to tell them I wasn't mister, I was Miss! But, after a few years, that was accepted.'

After the death of J.J. Bergin in 1958, Sean O'Farrell was appointed Managing Director of the NPA, and Anna May worked alongside him until his death in 1972. The following year, she took over the reins herself and was also secretary of the association.

She recalled the meeting when she was elected to the position and told *The Late Late Show*, 'I will never forget it! I was secretary, and I was recording the minutes of the meeting and I can assure you that when my name came up and when I was proposed and seconded and the lot, my pen refused to write. It was a complete surprise to me. Mind you, I came home and I realised that I had accepted responsibility and I only wished, for a moment, that the meeting was back again so that I could say no.'

'I'd have some convincing to tell them I wasn't mister, I was miss!'

*Sean O'Farrell and Anna May McHugh at the opening of
the Championships, 1969*

But she didn't hand it back, and she's 'not sorry'.

It was a sunny autumn day when I arrived at NPA headquarters in Fallaghmore, County Laois to meet Anna May McHugh and her daughter Anna Marie. It's a warm, friendly place and more like visiting the McHughs at home than the headquarters of Europe's largest agricultural show.

Over a cup of tea, I realise that Anna May has a phenomenal memory as she reels off dates and ploughmen and their achievements, hopping from county to county and around the venues, past and present. Over the years, she has met the great and the good and when she met Pope John XXIII, he told her that when he was a schoolboy, he remembered climbing over a wall and stealing apples!

Anna May married John McHugh in 1966. John died in January 2007, but Anna May is a woman of strong faith. 'God is in my life in a very strong way,' she told Gay Byrne when he interviewed her on *The Meaning of Life*. 'God, I feel, is with me everywhere I go. I would ask God for favours. I would pray for God to cure people and I don't know how I would survive if I did not have God in my life. I believe He is there to guide me.' She also said she looks forward to meeting John and her family again in Heaven.

The couple were complementary, John minding their two children, D.J. and Anna Marie when their mother was away at international ploughing matches. He accepted her work. Anna Marie says her dad was 'before his time' and, indeed, Anna May was at the World Championships in New Zealand when her daughter made her Confirmation. But she also remembers the presents were great when her mum returned. 'Mam wasn't there and people felt sorry for us. We got to go into the convent, we were wined and dined and

Anna May and her daughter Anna Marie McHugh

235

brought to a hotel for lunch and then got extra presents when Mam came home!'

Today Anna Marie works alongside her mother as an Assistant Managing Director of the NPA and also holds the post of Secretary of the World Ploughing Association (of which her mother is a board member).

Anna May is a much-loved household name, but she has also attracted the admiration of presidents. President Mary McAleese paid tribute to her, saying, 'Under her leadership, the National Ploughing Association has turned the National Championships into a major event in Irish agriculture. While ploughing remains the main focal point, the National Championships now cover not only all sectors of agriculture but also the social and cultural aspects of rural life.'

Anna May has been the recipient of a string of honours over the years including, in 2015, being named as Officier de l'Ordre du Mérite Agricole. The French Ambassador, Jean Pierre Thébault, presented her with the award at the 2015 Championships. He said there was no more fitting recipient of such an honour than Anna May. In fact she 'exemplified' it since 'almost from the time she was born, she was thinking, acting and working for agriculture'. The *Irish Independent* covered the event, reporting that former Rural Affairs Minister, Éamon Ó Cuív, described Anna May as 'extraordinary'. 'She's like a great hostess here and to see her you would never think she was under any pressure even though she's carrying such a workload,' he said.

Anna May has also been the recipient of a number of other awards, including Laois Person of the Year, the Millennium Laois Person of the Year, the Rehab People of the Year Award,

'Almost from the time she was born, she was thinking, acting and working for agriculture.'

*Anna May McHugh receives the Lifetime Achievement Award
at the FBD Women and Agriculture Awards*

the *Irish Tatler* Woman of the Year, the FBD Women and
Agriculture Awards – Lifetime Achievement Award, Veuve
Clicquot Business Woman of the Year Award, the *Irish
Times* Trailblazers Award 2014, and two honorary doctorate
degrees.

'God speed
the plough!'

As she often says herself, 'God speed the plough!'

IRISH FARMERS JOURNAL farmersjournal.ie

17 September 2016

PLOUGHING

2016

FREE

Inside this week's issue

NATIONAL PLOUGHING CHAMPIONSHIPS

Essential 96-page guide to the National Ploughing Championships, including:

➤ Free site map
➤ Farm machinery

➤ List of exhibitors
➤ Livestock exhibits

➕ PLOUGHING EXTRA IN THE MACHINERY SECTION

8. Ploughing and the Farmers' Bible

The *Irish Farmers Journal* began publishing in 1948 and, since then, it has become the bible for farmers and ploughing enthusiasts alike. It has grown from very modest coverage of the National Ploughing Championships to today's full-blown supplements on every aspect of the event, with hundreds of photographs and detailed analyses.

But to put this in an historic context, the *Journal* was published in the year following rural electrification, the biggest transformation of Ireland's rural communities and a project that gave light and heat and energy to Irish farms. It was just seventeen years after the first Nationals, which were still a small event and didn't warrant any front-page coverage.

The earliest front-page mention I found was in 1963, a small corner picture captioned 'Spectators at the National Ploughing Championships in Athenry formed this study in intentness as they watched the critical stage of closing a furrow'. But the reportage was direct, even to the extent that in 1961, Noel O'Reilly, writing about that year's event in Killarney, said he was 'a bit disappointed' with the demonstrations. 'We hear a lot nowadays about the value of the live demonstration and its appeal over the static display on the showground, but in Killarney where there was an extensive gathering of farmers and potential purchasers, little was done to attract them to the demonstration plots, or to hold their interest there.'

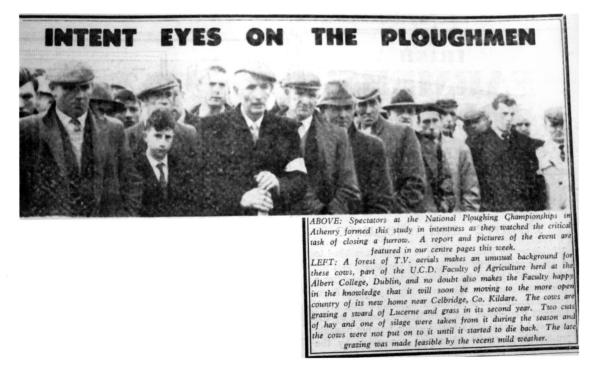

ABOVE: Spectators at the National Ploughing Championships in Athenry formed this study in intentness as they watched the critical task of closing a furrow. A report and pictures of the event are featured in our centre pages this week.

LEFT: A forest of T.V. aerials makes an unusual background for these cows, part of the U.C.D. Faculty of Agriculture herd at the Albert College, Dublin, and no doubt also makes the Faculty happy in the knowledge that it will soon be moving to the more open country of its new home near Celbridge, Co. Kildare. The cows are grazing a sward of Lucerne and grass in its second year. Two cuts of hay and one of silage were taken from it during the season and the cows were not put on to it until it started to die back. The late grazing was made feasible by the recent mild weather.

The *Journal* was also concerned about the Irish ploughing 'style' and, in 1960, said it was hoped to give our ten contestants 'some training in world ploughing styles which, as this year's competitors learned at Rome, is very much different to our national standards'.

Rural electrification

On 1 January 1948, the *Irish Times* columnist Quidnunc reviewed some of the events of the previous twelve months. He concluded, 'How many of these things will be remembered in, say, 2047? I dare swear that if any event is recorded in the history books, it will be none of those I have mentioned.'

What the writer guessed would be most significant was the fact that in 1947, 'Somebody – I cannot remember who – switched on the lights in some village – I cannot remember where – and rural electrification took her bow. And if that does not mean more to the country than all the rest of the year's events put together, I shall be very surprised indeed.'

The first pole in the first phase of rural electrification went up on 5 November 1946, at Kilsallaghan, in North County Dublin. The first lights of the scheme were switched on at Oldtown, County Dublin, in January 1947. At the time more than 400,000 homes in rural Ireland had no electricity.

Quidnunc was entirely correct in his prediction about the importance of rural electrification. It transformed the lives of hundreds of thousands of people, made farming and household tasks less challenging and helped to reduce social isolation.

In 1964, it lamented the drop in the number of entries in the horse classes which were 'gradually declining' and wondered who would take the place of 'the long reigning champions from Kerry, Wicklow and Carlow' when they retired. And the same year, as it previewed the National Championships to be held in Danesfort, County Kilkenny, the *Journal* noted that 'the success of World Ploughing champion, Charlie Keegan, fresh from his prestigious win in Vienna, adds a great fillip to the All-Ireland Ploughing Championships to be held next week at Edward

Walsh's farm'. There was much to look forward to because 'we have been informed by NPA Secretary, Miss Brennan, that there is a record entry of ploughmen in all classes with all the old hands competing'.

Also in 1964, there was a new Intermediate competition that Sean O'Farrell, Managing Director of the NPA, said was 'designed to whet and hold the interest of younger ploughmen who have proved their ability in the minor grades but very often became disheartened and lost interest when faced with the stiff competition of more experienced ploughmen when they moved out of that grade'.

Irish farmers' journal

THE VOICE OF IRELAND'S BIGGEST INDUSTRY
Vol. 32 No. 39 October 11, 1980. Price 25p.

Farm taxation
special supplement
Centre pages

Prospects for winter beef fattening
page 35

Colour at the ploughing championships

The first day of the national ploughing championships at Rockwell College brought thunder and lightning, hail and gale force winds. Adding colour with their headgear to one of the few bright intervals were (l. to r.) Cork visitors: Ben Withers, Freemount; John Brosnan, Liscarroll; Oliver Casey, and Denis Kearney, Kilbrin. See page 19.
Photo Steve Treacy.

In October 1980, the front page photograph by Steve Treacy is of four men in bobble hats captioned, 'Colour at the ploughing championships: The first day of the ploughing championships at Rockwell College brought thunder and lightning, hail and gale force winds.

The *Journal* has kept and preserved a wonderful archive of all their publications over the years and I spent a few glorious days covered in dust and that very special old newspaper smell which promises hidden treasures to those who persevere.

Leaving the vault behind, I went to meet Matt Dempsey who was the editor of the *Journal* for almost a quarter of a century. Today, he is chairman, columnist, and also a former chairman and former president of the RDS. He understands, perhaps more than most, how the progress and expansion of the National Ploughing Association was shaped by the RDS.

> The crucial change came in 1979–1980. The RDS Spring Show was the main shop window by a mile for the agricultural machinery and the FTMTA [the Farm Tractor and Machinery Trade Association] wanted some kind of demonstration area at the Spring Show and, in my view, I think the RDS handled it badly. Instead of trying to facilitate them, they simply said that it wasn't an option and dismissed them. So, there was an actual break in the relationship between the FTMTA and the RDS. Enter Anna May and she filled that void.

Matt remembers one of the first ploughing matches he attended, in Finglas in 1971.

> Primarily, there were a few bits and pieces of machinery in the corner of a field and a few guys ploughing and that was it. There

was nothing much there, whatsoever! So it was that row between the RDS and the FTMTA that really set her up, that break in the relationship. Then, the FTMTA started doing their own show, but they only did it on a biennial basis, and that's the one that's now in Punchestown. But it's no real threat to the Ploughing. I suppose that was the crucial cross-over stage.

The RDS reached a peak of attendance of something like 240,000 in the 1978 Spring Show. It was absolutely enormous and everything was bulging at the seams, the economy was buoyant, it was just after the 1977 general election, agriculture was at the end of the transition period of EU membership – prices were at their peak in real terms, farm incomes were very high. That was the absolute peak. Then came the 1979–1980 disagreements.

By 1987, for the first time, the ploughing supplement in the *Journal* was larger than the Spring Show supplement. 'That was a continuation of a trend of the growth of the Ploughing and the erosion of the position of the RDS.'

In 1988, Matt took over as editor of the *Journal*. 'We had come through some extremely difficult years of low economic growth and high inflation – those that had borrowed heavily suffered really badly. Interest rates in the 1980–1986 period peaked to 20 per cent, it's inconceivable what they were in today's terms.'

> '**By 1987, the ploughing supplement in the *Journal* was larger than the Spring Show supplement.**'

Matt says that when he took over the paper, he was 'very comfortable' with its ethos, which was to enhance the competitiveness of Irish agriculture and the well-being of those involved in the sector. 'I loved it, to be honest with you. I have a farm near Maynooth, the farm I grew up on. My dad was a farmer and I helped on the farm as a kid.'

> '**I loved it, to be honest with you.**'

Matt Dempsey, former editor of the Irish Farmers Journal

In his twenty-five years as editor, there have been major changes in agriculture.

> Probably it's that shift from the conviction that the market should return the full costs of production to where we're essentially on world market prices, with some payments from Brussels, to make up the difference in competitiveness between us and some other parts of the world, and also for the extra imposition of environmental obligations on farmers. It's the environmental oversight, the necessity for literacy in the broadest sense, whether it's implementing regulations, claiming payments that are due with complicated forms, so it's that intrusion of Brussels and other policy into the normal daily existence of farmers.

Also, Matt says that farmers, instead of diversifying their enterprise, have tended to diversify their use of time so there has been very little movement. 'They're holding on to the land but they're seeing what other off-farm opportunities there are, either for their families or within community distance for themselves and their wives. In the main, it's off-farm employment, some of it very good, some of it not so good, depending on the level of education, qualifications and location as to where they are in the country. So, I think there has been a revolution from that point of view in farmers' perception of the capacity of the farm to provide an acceptable living today.'

'They're holding on to the land but they're seeing what other off-farm opportunities there are.'

Then there's the flip side of mechanisation on the farm which has meant that one man can quite comfortably cope now with 600 or 700 arable acres. One man can quite easily handle 300–400 cattle.

Modern machinery is incredibly powerful. So the number of farm employees has shrunk hugely. This, in my view, has created a social dilemma especially with wives and spouses tending to have off-farm jobs. And that is partly too, of course, why the Ploughing has been so successful, because everybody takes off for the three days, it's such a sectoral family gathering place. Ploughing has tapped a chord and supplies a need in a social sense of a sectoral get-together. Before this, someone stood up on the back of a tractor and trailer and declared it open, now the whole opening ceremony is something like the Olympic Games between the President of Ireland and every ambassador that ever drew breath!

I think Anna May also tapped in to this political view, that politicians like to be associated with some farming event, so when it became established at all as an acceptable political gathering place, when the numbers came, politicians were going to come to see and be seen. The Ploughing is now recognised as the essential meeting place!

During my travels around Ireland, I have seen many a copy of the *Journal* on many a kitchen table. And its circulation reflects that, selling to over 247,000 readers. In addition, the *Country Living* magazine is the biggest-selling women's magazine in Ireland.

But the *Irish Farmers Journal* is more than a farming newspaper, it sits around the house for a week at a time while the family comes to grips with the news, the farm prices and the small ads. In a strange way, it *is* part of the family.

Appendix

1973	John Tracey	2003	John Tracey	2013	Eamonn Tracey
1974	John Tracey	2004	Eamonn Tracey	2014	Eamonn Tracey
1980	John Tracey	2005	John Tracey	2015	Eamonn Tracey
1988	James Murphy	2006	Eamonn Tracey	2016	Eamonn Tracey

Overall World Champions

2014   Eamonn Tracey (Conventional)
2015   Eamonn Tracey (Conventional)

CORK

Senior Horse Plough Class

1971	Jerry Horgan	1987	Thady Kelleher	1997	Thady Kelleher
1972	Jerry Horgan	1992	Murty Fitzgerald	1999	Thady Kelleher
1974	John Halpin	1995	J.J. Delaney	2005	John Sheehan
1975	Jerry Horgan	1996	Thady Kelleher	2006	John Sheehan
1977	J.J. Egan				

Senior Conventional Plough Class

1978   Cyril Dineen
1999   Charles Bateman

Senior Reversible Plough Class

2009   Liam O'Driscoll
2011   Jeremiah Coakley

Queen of the Plough

1974	Lilian Stanley	1986	Marion Stanley	1990	Elizabeth Lynch
1980	Lillian Stanley	1987	Marion Stanley	1992	Marion Stanley
1985	Marion Stanley	1988	Elizabeth Lynch		

International House Ploughing Champion

1984   Thady Kelleher
1996   Thady Kelleher

Competitors at the World Ploughing Championships

1969	Cyril Dineen	1991	Charles Bateman	2000	Charles Bateman
1979	Cyril Dineen	1991	Jackie O'Driscoll	2001	Charles Bateman
1983	Jeremiah Coakley	1994	Jackie O'Driscoll	2006	Liam O'Driscoll
1985	Jeremiah Coakley	1995	Charles Bateman	2009	Liam O'Driscoll
1986	Jeremiah Coakley	1996	Jackie O'Driscoll	2010	Liam O'Driscoll
1989	Charles Bateman	1997	Mervyn Buttiner	2012	Jeremiah Coakley
1990	Charles Bateman				

DUBLIN

Senior Conventional Plough Class

1954   William R. Murphy
1958   William R. Murphy
1961   William R. Murphy
1968   William R. Murphy
1970   William R. Murphy

Queen of the Plough

1971   Maura Murphy
1972   Maura Murphy

Competitors at the World Ploughing Championships

1954   William R. Murphy

1961 William R. Murphy
1963 William R. Murphy

GALWAY

Senior Horse Plough Class

1978 Thomas Reilly	1988 Thomas Reilly	1994 Joe Fahy
1981 Thomas Reilly	1989 Thomas Reilly	2000 Joe Fahy
1982 Thomas Reilly	1990 Joe Fahy	2014 Gerard Reilly
1983 Thomas Reilly	1991 Thomas Reilly	2016 Gerard Reilly
1985 Thomas Reilly	1993 Thomas Reilly	

Queen of the Plough

1957 Eileen Duffy

KERRY

Senior Horse Plough Class

1952 Patrick O' Mahony
1954 J.J. Egan
1958 J.J. Egan
1960 Patrick O' Mahony
1961 J.J. Egan
1963 Patrick O' Mahony

Senior Conventional Plough Class

1951 J.P. Shanahan
1957 Con Slattery

Queen of the Plough

1955 Anna Mai Donegan
1956 Anna Mai Donegan
1959 Mary Shanahan
1960 Mary Shanahan

KILKENNY

Senior Horse Plough Class

1943 Patrick Nolan
1945 Peter Murphy

Senior Conventional Plough Class

1959 Michael Muldowney	1965 Michael Muldowney	1972 William Ryan
1960 Richard Mullanny	1966 Michael Muldowney	1974 William Ryan
1962 Michael Muldowney	1969 Michael Muldowney	1982 William Ryan
1963 Michael Muldowney		

Senior Reversible Plough Class

2000 Brian Ireland
2005 Brian Ireland

Queen of the Plough

1958 Peggy Mulally
1970 Mary Ryan
1973 Mary Ryan
1981 Alice Murphy
2006 Lisa Hartley

Competitors at the World Ploughing Championships

1960 Michael Muldowney	1975 William Ryan	1981 William Ryan

| 1961 | Michael Muldowney | 1978 | William Ryan | 1983 | William Ryan |
| 1963 | Michael Muldowney | | | | |

LAOIS

Senior Horse Plough Class

1948 Liam O'Connor

Senior Conventional Plough Class

1986 Liam Rohan

Queen of the Plough

1961 Eileen Brennan
1963 Eileen Brennan
1964 Eileen Brennan
2008 Anna Marie McHugh

Competitors at the World Ploughing Championships

1982 Liam Rohan
1985 Liam Rohan
1987 Liam Rohan
1988 Liam Rohan

LOUTH

Senior Horse Plough Class

1998	Gerry King	2007	Gerry King	2011	Gerry King
2002	Gerry King	2008	Gerry King	2012	Gerry King
2003	Gerry King	2009	Gerry King	2013	Gerry King
2004	Gerry King	2010	Gerry King	2015	Gerry King

Senior Conventional Plough Class

1945 John Butterly
1950 John Butterly
1953 Andrew Rogers
1956 Henry Rowley

Competitors at the World Ploughing Championships

1953 Thomas McDonnell

MONAGHAN

Queen of the Plough

| 1983 | Elizabeth McCaul | 2011 | Joanne Deery | 2014 | Joanne Deery |
| 1984 | Elizabeth McCaul | 2013 | Joanne Deery | 2015 | Joanne Deery |

NORTHERN IRELAND

Senior Conventional Plough Class

2002 David Wright

Senior Reversible Plough Class

1995 George Murphy
1997 Adrian Jameson
2004 Thomas Cochrane
2008 David Wright

Overall World Champions

| 1954 | Hugh Barr | 1986 | Desmond Wright |
| 1955 | Hugh Barr | 1997 | Thomas Cochrane |

1956	Hugh Barr	2003	David Wright
1959	Hugh Barr	2007	David Gill
1984	Desmond Wright		

OFFALY
Queen of the Plough

1997 Fiona Claffey
2016 Laura Grant

TIPPERARY
Competitors at the World Ploughing Championships

2000 John Slattery

WATERFORD
Senior Reversible Plough Class

1992 Pat Barron

Queen of the Plough

1962 Angela Galgey

WEXFORD
Senior Horse Plough Class

1931	Ned Jones	1938	David O'Connor	1950	William Kehoe
1932	Michael Redmond	1939	Michael Redmond	1951	Michael Redmond
1933	Ned Jones	1940	Michael Redmond	1957	John Kent
1934	Michael Redmond	1941	William Kehoe	1969	John Kent
1936	Michael Redmond	1949	William Kehoe	1973	John Kent
1937	Ned Jones	1950	William Kehoe	1979	Michael Kinsella

Senior Conventional Plough Class

1955	William Kehoe	1988	Martin Kehoe	1995	Martin Kehoe
1964	Andrew Cullen	1989	Martin Kehoe	1996	Martin Kehoe
1976	Joseph Shanahan	1990	Martin Kehoe	1997	Martin Kehoe
1979	John Summers	1991	Martin Kehoe	1998	Martin Kehoe
1981	Martin Kehoe	1992	Martin Kehoe	2007	William J. Kehoe
1984	John Summers	1993	Martin Kehoe	2009	William J. Kehoe
1987	Martin Kehoe	1994	Martin Kehoe		

Senior Reversible Plough Class

1993	James Walsh	2002	Dan Donnelly	2012	John Whelan
1994	James Walsh	2003	Dan Donnelly	2013	John Whelan
1996	James Walsh	2006	John Whelan	2014	John Whelan
1998	James Walsh	2007	John Whelan	2015	John Whelan
1999	James Walsh	2010	John Whelan	2016	John Whelan

Queen of the Plough

1965	Bridget O'Connor	1991	Deirdre Barron	2005	Christine Kehoe
1966	Bridget O'Connor	1996	Deirdre Barron	2007	Michelle Kehoe
1967	Bridget O'Connor	2000	Michelle Kehoe	2009	Eleanor Kehoe
1969	Lillian Keating	2002	Michelle Kehoe	2010	Bernadette Nolan
1975	Majella Ffrench	2003	Michelle Kehoe	2012	Eleanor Kehoe
1976	Majella Ffrench				

Competitors at the World Ploughing Championships

1960	Andrew Cullen	1984	John Somers	2005	John Whelan

1964	Andrew Cullen	1986	Martin Kehoe	2007	John Whelan
1965	Andrew Cullen	1987	Martin Kehoe	2008	John Whelan
1967	James Murphy	1988	James Murphy	2008	William John Kehoe
1970	James Murphy	1989	Martin Kehoe	2010	William John Kehoe
1971	James Murphy	1992	Martin Kehoe	2011	John Whelan
1972	Joseph Shanahan	1993	Martin Kehoe	2013	John Whelan
1976	Joseph Shanahan	1994	Martin Kehoe	2014	John Whelan
1977	Joseph Shanahan	1995	Martin Kehoe	2015	John Whelan
1979	John Somers	1996	Martin Kehoe	2016	John Whelan
1980	John Somers	1998	Martin Kehoe	2015	John Whelan
1981	John Somers	1999	Martin Kehoe	2016	John Whelan
1982	Martin Kehoe	2003	William John Kehoe		

Overall World Champions

1994	Martin Kehoe
1995	Martin Kehoe
1999	Martin Kehoe
2013	John Whelan

WICKLOW

Senior Horse Plough Class

1935	Hugh Pierce	1964	Peter Byrne	2011	Gerry King
1946	John Halpin	1964	John Halpin & Peter Byrne	2012	Gerry King
1947	John Halpin	2009	Gerry King	2013	Gerry King
1959	John Halpin	2010	Gerry King	2015	Gerry King

Senior Conventional Plough Class

1952	William Woodruff
1971	Charles Keegan

Queen of the Plough

1957	Muriel Sutton	1968	Betty Williams	1978	Pauline O'Toole
1959	Muriel Sutton	1977	Gretta O'Toole		

Competitors at the World Ploughing Championships

1953	Ronald Sheane	1962	Charles Keegan	1966	Charles Keegan
1954	Ronald Sheane	1964	Charles Keegan		

Overall World Champions

1964	Charles Keegan

Permissions
Acknowledgements

The author and publisher would like to thank the estate of J.J. Bergin for permission to reproduce 'The Song of the Plough', and the following for permission to use inside photographs in *A Ploughing People*:

© Agriland: 213T; © Barr Family: 183, 184, 248CR; © Bergin Family: 4, 5, 7 (both images), 12, 14, 15, 22 (both images), 23 (both images), 25 (both images), 31, 33, 35, 37, 118, 243, 244; © Brandon Family: 66; © Brennan Family: 70; © Clissmann Family: 101, 103; © Trina Connolly: 248BR; © Cotter Family: 122 (both images); © Deery Family: 204R; © Embrace FARM: 62; © Fahy Family: 71, 74; © Hamilton Family: 87B; © Hurley Family: 88; © Imagno/Getty Images: vi; © *Irish Examiner*: ii, 32, 41, 132, 135, 138, 148, 157, 161, 209BL, 217; © *Irish Farmers Journal*: 8, 42, 50, 53, 55, 59, 82, 113, 170, 181, 185, 195, 204L, 207, 215, 229, 234, 238, 240, 242, 247, 248TL, 248TR, 248TL, 248TR, 248C; © *Irish Farmers Journal*/Bergin Family: 219T; © *Irish Farmers Journal*/Jack Caffrey: xii, 57,

Acknowledgements

I have loved researching and writing this book and I want to thank all the ploughing people who welcomed me into their families for a short time, who shared their wonderful stories of days long ago, memories of fathers and grandfathers, of the champion ploughmen who now work the kinder fields of Heaven. I hope I have done their stories justice.

Thank you too to the local Ploughing communities who shared their own publications with me and who provided photographs for this book; to all the staff in the *Irish Farmers' Journal* who allowed me into their archives, and the staff of Teagasc at Oak Park. Thank you to the officers of the National Ploughing Association for all the background detail and sharing the contacts.

I hope your stories of rural Ireland and the ploughing will ensure that the children of a new generation will remember your traditions and your triumphs. The memories are there, in a well-turned sod on a frosty morning, or in the gentle spring breezes where hearts remember.

I also want to thank my publishers Hachette Books Ireland: my patient and imaginative editor Ciara Considine, copy-editor and picture researcher Claire Rourke, Joanna Smyth and all the team there.

God Speed the Plough!